BETTER *Science* THROUGH *Safety*

BETTER *Science* THROUGH *Safety*

JACK A. GERLOVICH
GARY E. DOWNS

Iowa State University Press
AMES, IOWA

© 1981 The Iowa State University Press
All rights reserved

Composed and printed by
The Iowa State University Press
Ames, Iowa 50010

No part of this publication may be reproduced,
stored in a retrieval system, or transmitted,
in any form or by any means, electronic, mechanical,
photocopying, recording, or otherwise, without the
prior written permission of the publisher.

First edition, 1981

International Standard Book Number: 0-8138-1780-3

CONTENTS

	Preface	vii
	Acknowledgments	ix
1/	Introduction and Statistical Background	3
2/	Legal Liability	9
3/	Eye Protection and Eye Care	19
4/	Safety in Biology Settings	25
5/	Safety in Chemistry Settings	41
6/	Safety in Physics Settings	77
7/	Field Activities	81
8/	Student Research (Projects)	85
9/	Physical Plant and Facilities	87
10/	Science Safety for Handicapped Students	95
11/	Accident/Incident Reporting Systems	101
	Appendices	
	A/ Checklists: Legal Liability, 107; Eye Protection, 107; Eye Care, 108; Biology Settings, 108; Chemistry Settings, 110; Physics Settings, 111; Field Activities, 112; Student Research, 112; Physical Plant and Facilities, 112; Handicapped Students, 113; Accident/Incident Reporting, 113.	
	B/ Eye Protectors	114
	C/ Explosion-safe Refrigerators	116
	D/ Safety Equipment	122
	E/ First Aid	131
	Index	143

PREFACE

Some level of risk is inherent in all science activities. The problem is to determine an acceptable level of risk for all planned activities contained in the science curriculum.

As safety philosopher Herman H. Horne has stated, "Life at its best is taking risks for the things worthwhile. The good life is adventuring in the creation of values. Safety has its rightful place when no greater value is at stake for which a risk should be taken." Safety enables us to choose between experiences that are unproductive or even foolish and those that enrich our lives and make them worthwhile.

The Council of State Science Supervisors believes that safety practices are learned and habits are formed by following models presented by others. Therefore, it is important that all science teachers understand the implication their safety practices have for students who will learn from them.

It is with this philosophy in mind that we present *Better Science Through Safety* to teachers everywhere and urge its use.

Jack A. Gerlovich
Gary E. Downs

ACKNOWLEDGMENTS

The Iowa Department of Public Instruction commends the following members of the Science Safety Task Force for their foresight in developing a practical and valuable science safety manual. In addition, the Department of Public Instruction gives special thanks to the National Science Foundation for helping disseminate this manual via a grant through Iowa State University, NSF Grant No. SER 7909642, Gary E. Downs, Director.

Any opinions, findings, conclusions or recommendations expressed (herein) are those of the authors and do not necessarily reflect the views of the National Science Foundation.

Science Safety Task Force

Jack A. Gerlovich (Co-Chair)
Science Consultant
Department of Public Instruction

Frank W. Starr (Co-Chair)
Science Supervisor
Waterloo, Iowa, Community Schools

Clifford G. McCollum
Dean, College of Natural Sciences
University of Northern Iowa

Burgess Shriver
Director, Math/Science
Des Moines Area Community College

Gil Hewett
Science Consultant
Area Education Agency 7

Thomas Scott
Chemistry Teacher
Lincoln High School, Des Moines

Marvin Van Sickle
Consultant, Traffic Safety Education
Department of Public Instruction

Milton Wilson
Consultant, Plant Facilities Unit
Department of Public Instruction

Elwood Garlock
Science Teacher, Taft Junior High School
Cedar Rapids, Iowa

William Peterson
Science Consultant
Western Hills Area Education Agency

Special thanks to the following individuals who served in consultative and/or coordinating capacities.

Gary E. Downs, Director
National Science Foundation Grant
Science Education
Iowa State University

Joseph R. Songer
Biohazard Control Office
National Animal Disease Center
Ames, Iowa

Joseph W. Klinsky
Environmental Health, Safety
Iowa State University

Jack A. Beno
Safety Education
Iowa State University

Walter L. Hetzel
Attorney at Law
Ames, Iowa

Norman H. Anderson--CSP
Risk Improvement Dept.
Employers Mutual Company
Des Moines, Iowa

Albert M. Sherick
Industrial Education
Iowa State University

John H. Thomas
Optometrist
Valley Optometric Center
Des Moines, Iowa

Alice F. Suroski
Special Education
University of Northern Iowa

Frank Vilmain
Physics
University of Northern Iowa

Harold Rathert
Science Coordinator
Des Moines Independent Schools

BETTER *Science* THROUGH *Safety*

Introduction and Statistical Background

Three major components are found in effective school safety programs: safety management, safety education, and safety services.

Safety management includes organizing, planning, budgeting, staffing, implementing, recording, coordinating, evaluating, and interpreting programs. Characteristics of a well-managed program include: (1) a specific program plan with stated goals and sound policies derived from realistic needs; (2) qualified, interested, and experienced coordinators who make efficient use of personnel, facilities, and resources; (3) cooperative and capable school personnel; (4) active communications within the school district and between the school and community; and (5) an ongoing process of evaluation and program improvement.

Safety education involves provisions of meaningful learning experiences for students and inservice education for school personnel. Successful functioning of the instructional program is enhanced by: (1) enthusiastic school personnel, including administrators and teachers who make effective use of safety education preparation received from inservice experiences and/or college or university courses; (2) a well-planned curriculum; (3) up-to-date and effective methods, techniques, equipment, and materials; and (4) thorough procedures for improving the program.

Safety services are best delivered in schools that are designed and built for optimal safety, where there is good inspection and maintenance of the total school environment; and through the establishment of adequate procedures for supervision of student activity and the handling of emergencies. To be effective, this component should include: (1) specific operating procedures, (2) appropriately scheduled inspections and reporting, (3) qualified personnel, and (4) the cooperation and assistance of the school community. These goals tend to be defensive. They should not be so. Teachers have a unique opportunity to impart some of these previously omitted aspects of safety into the student's formal education.

SCHOOL PLANT AND PERSONNEL SERVICES

The services of a school safety program stress safety on the school site. A safe school environment involves physical and engineering considerations and administrative actions designed to identify, eliminate, or minimize hazards by

---

Portions of this chapter were adapted from a statement prepared by the 6th National Conference on Safety Education, December 3-8, 1978, Central Missouri State University, Warrensburg, Missouri.

means of inspections of buildings, grounds, and equipment at regular intervals and the taking of proper remedial action. The following are examples of desirable characteristics of effective school plant and personnel services:

-- Funds are budgeted for safety services.
-- An accident reporting and record system is in operation which includes investigation, analysis, and utilization of results to reduce or eliminate hazards, apply corrective measures, and improve instruction.
-- A personal emergency information file is maintained for students and school personnel.
-- Maintenance programs include safety inspections which result in compliance with local, state, and federal standards.
-- A plan is in effect for safe routing of students in and around the school plant and to and from school.
-- Written policies govern the use of facilities and all student activities.
-- Security procedures provide for the safety of students, school personnel, and visitors.
-- Up-to-date and accurate information is available regarding accident liability and insurance.
-- Students help plan the safety program and participate in it by serving as officers and members of school safety patrols, committees, and clubs.
-- Drills based on written procedures, designed to prepare students and school personnel for proper action in natural and manmade emergencies, are conducted on a systematic basis.

COMMUNICATIONS

Effective communications are vital to the success of school safety programs. Communications can assist a school district in:

-- Establishing the organizational and managerial aspects of the safety program.
-- Promoting participation in the safety program by students, school personnel, and community representatives.
-- Facilitating communications among students, school personnel, parents, community leaders, district and state school safety specialists, resource personnel from state agencies, and college and university personnel by:
    ... Providing speakers for interpretation of the school safety program to civic and service organizations, business, industry, and to the community.
    ... Inviting the community to school-sponsored safety activities.
    ... Utilizing resource people and parents in classroom safety education activities.
    ... Encouraging field trips and planned excursions.
    ... Placing school representatives, including students, on community-wide committees concerned with safety.
    ... Communicating information to parents concerning holiday and seasonal safety, accident statistics and research findings, community hazards, program progress, and suggested home and recreational safety activities.
    ... Securing advice and assistance from an advisory committee composed of students and the following: police, fire, and health agencies; motor clubs; business and industrial groups; civic and service organizations; parent/teacher organizations; district administrators; media;

Introduction and Statistical Background

and college and university personnel in efforts to identify, eliminate, or minimize school and community hazards.

A SCIENCE SAFETY PROGRAM

A review of the legal cases involving suits against teachers might lead the unsuspecting to believe that in recent years few accidents have occurred in the area of science in schools. Accidents have occurred; however, in most cases, insurance companies have settled claims out of court.

Due to the general lack of national data describing accidents in science classes, information available in Iowa was compiled and utilized as a representative data base.

During the 1977-78 and 1978-79 academic years Employers Mutual Companies (EMC) of Des Moines provided liability insurance for approximately 45 percent of Iowa's 449 school districts. During this time, 8,476 student injury reports were reported to EMC from 88 educational units. Table 1.1 indicates the general nature of all student injuries reported and the occurrence percentages.

Table 1.1. GENERAL NATURE OF ACCIDENTS

	%
Bruises	31
Sprains/Strains	23
Cuts	19
Fractures	13
Scratches	5
Dislocation	2
Concussion	2
Burn, Scald (heat)	1
Multiple Injuries	1

Upon closer analysis, the data indicate that the greatest number of accidents involve injury to the face. Table 1.2 indicates injury by body part.

Table 1.2. ACCIDENTS INVOLVING BODILY INJURY

	%
Face	19
Leg	18
Hand/Fingers	17
Head	13
Arm	9
Eyes	4
Foot/Toes	4
Back	3
Shoulders	3
Neck	2
Hips	2
Multiple Parts	1
Chest	1

When student injury reports were coded, an approximate severity level was entered for each report. The severity analysis is summarized in Table 1.3.

Table 1.3. ACCIDENT SEVERITY

	%
First aid not required	10
First aid	43
Minor Medical	36
Medical	10
Major Medical	1
Fatality	1 case

Minor Medical--X ray, 7 stitches or less
Medical--without hospitalization, 8 or more stitches
Major Medical--hospitalization required

A total of 94 claims involving student injury resulted in dollar payments or reserves of approximately $400,000 in the time period July 1976 through January 1979. Camping and hiking trips, falls from bleachers, physical education and vocational shop accidents were responsible for most of these losses. Science and chemistry classes involved only one accident in this group which was a rocket launching experiment that failed; a young girl received slight burns.

The 8,476 incident reports can be separated by activity type as shown in Table 1.4.

Table 1.4. ALL ACTIVITIES

		%
Recess	2,534	30
Gym	1,846	22
Classroom Activity	995	12
Athletics Practice	879	10
Athletics Org. Game/Competition	614	7
Horseplay	490	6
Hallway Movement	469	6
Activity not on School Grounds	252	3
Fighting	208	3
Leaving/Exiting School Bldg.	64	1
Entering School Bldg.	61	1
Insufficient Data/Undetermined	44	1

Science incidents are included in the 995 reports in the Classroom Category. General Classroom reports were indicated in 324 of the accidents, and 671 of these accident reports are further separated as shown in Table 1.5.

Table 1.5. GENERAL CLASSROOM ACCIDENTS

		%
Lunch/Free Time	140	21
Art/Ind. Arts	124	18
Woodworking	104	16
Metalworking	73	11
Musical/Play	51	8
Home Ec./Foods	44	7
General Science	40	6
Auto Mechanics	27	4
Walking	15	2
Welding	15	2
Chemistry	15	2
Biology	12	2
Playing	11	2

Table 1.6 summarizes the nature of accident in the General Science, Chemistry, and Biology classes from Table 1.5.

Table 1.6. NATURE OF ACCIDENTS IN SCIENCE

	%
Glass	21
Chemicals	18
Animals	13
Another Person Involved	10
Laboratory Utensils	6
Metal Item	4
Thermometer	4

Close analysis of the above two years of data indicates that there have not yet been serious accidents in science classes. Although the conditions may present potential hazards, these need not result in serious student injuries when properly supervised.

A well-designed and fully implemented science safety program will contribute to the total school safety program. Science teachers can meet the requirements of such a program by:

-- Establishing and maintaining a safe environment.
-- Providing appropriate first aid.
-- Providing learning experiences which offer students opportunities to acquire positive behaviors for safe living.
-- Guiding student development of perceptual-motor skills for proper, safe, and efficient use of facilities and equipment.
-- Developing in students the ability to assess risks and make and implement appropriate decisions.
-- Assisting other school personnel in identifying students who are involved in repeated accidents and taking appropriate remedial actions.
-- Assisting other school personnel in identifying students who, because of special needs, may require special instruction, facilities, and materials.
-- Providing leadership and supervision for co-curricular safety activities.
-- Participating fully in a school-wide accident reporting system.
-- Serving on school building safety committees.

Students should be encouraged to help plan the science safety program and to participate in it by serving as members of school and community safety committees and safety organizations, in addition to being safety-aware members of science classes.

2

Legal Liability

Much of the information provided in this chapter is based on the *School Laws of Iowa*; however, it is representative of state laws dealing with education throughout the United States.

Teachers, as well as the state, the school district, the school board, and the school administration can incur legal liability in a number of ways; however, one only is of interest here, negligent behavior.

In all states it is the teacher, himself, who is legally responsible for the safety of his pupils. However, the courts have held that a teacher is liable for damages only if it can be proven that the teacher has failed to take "reasonable care" or has acted in an illegal manner. A teacher must foresee dangers, but only to the extent that any reasonably prudent person would. He may punish a child, within the legal limits set by his state, as long as the punishment is humane, is not excessive, and provided the pupil knows why he is being punished. In addition, a teacher must perform his assigned duties if he is to avoid being censured by the school district in which he is employed.[1]

NEGLIGENCE IN TORT LAW

Negligence in the eyes of the law may be defined as conduct that falls below a standard of performance established by law to protect others against an unreasonable risk of harm.

Whenever one is injured, it does not necessarily follow that he will collect from another. Courts must first decide the cause of the injury. In making this determination, courts usually pose four questions: (1) Did one owe a care of duty to another? (2) Did one fail to exercise that care of duty? (3) Was there an accident in which a person was injured? And, (4) Was the failure to exercise that care of duty the proximate cause of the injury? If it can be shown that failure to exercise the care of duty was the direct cause of the injury, then the defendant may be liable. Relief may or may not involve a monetary award.[2]

---

1. B. W. Brown and W. R. Brown, *Science Teaching and the Law*. Washington, D.C.: National Science Teachers Association, 1969.
2. H. C. Hudgins, Jr., "Tort Liability," in the Yearbook of School Law 1976, P. K. Piele, ed. Topeka, Kansas: National Organization on Legal Problems of Education, 1976.

DUE CARE

Before liability can be incurred by a teacher, a determination must be made whether that individual exercised due care. It must be decided if the defendant reasonably foresaw a potential problem, or should have seen the potentiality of a problem, and took necessary and prudent measures to prevent it. If he failed to exercise that standard of due care, and if the failure was the proximate cause of the injury, then the plaintiff may recover for damages.[3]

REASONABLE AND PRUDENT JUDGMENT

If the standard of care has not been specifically established by statute, the actions or inactions of an individual will be measured against what a hypothetical, reasonably prudent individual would have done under the same circumstances.[4] Obviously there can be legitimate and complex questions regarding the course of action which a prudent person can take under a given set of circumstances.

One important aspect of the conduct of the reasonable person is anticipation. A reasonable person is expected to be aware of the foibles of human nature and be able to anticipate where difficulties might arise. Thus a pedestrian may not step blindly in front of a moving vehicle expecting the driver to stop, and a teacher may not direct a student to perform a dangerous experiment without giving adequate instruction and supervision. The reasonable teacher must be able to anticipate the common ordinary events and, in some cases, even the extraordinary. (See Appendix A, p. 107.)

DUTIES OF THE TEACHER

The classroom teacher has three basic duties related to the legal concept of negligence. These are the duty of *instruction*, the duty of *supervision*, and the duty of *proper maintenance and upkeep of all equipment and supplies used by students*.

Science teachers have responsibility for all three duties. Students in classroom, field, or laboratory settings should not be allowed to engage in an activity without first receiving complete instructions from the teacher. The teachers should include in such instructions an explanation of the basic procedure involved, some suggestions on conduct while performing the activity, and the identification and clarification of any risks involved.

One of the most frequent causes of an accident at school is the failure of personnel to instruct properly and supervise sufficiently. When the educator has been derelict in his duty and this dereliction is the proximate cause of the accident, then the injured party may recover in damages. On the other hand, failure to instruct and supervise that is not the direct cause of an accident will carry no liability.[5]

A defense often used in a tort suit is that of contributory negligence. It holds that the plaintiff was the direct cause of the injury and thus has no suit for recovery.[6]

School personnel cannot ensure the safety of another from physical defects. They are expected, however, to take reasonable precautions in inspect-

---

3. Ibid., p. 55.
4. S. R. Ripp, "The Tort Liability of the Classroom Teacher," *Akron Law Review* 9 (1975):19.
5. Hudgins, p. 58.
6. Ibid., p. 63.

ing the school premises, noting any dangerous conditions, and taking necessary and appropriate steps to correct them.[7]

In Iowa, although the teacher, as well as others affiliated with the school system may incur liability, even while utilizing reasonable and prudent judgment, governmental subdivisions would be required under the save harmless provision (613A.8) of Chapter 613.A of the Iowa Code to protect the teacher and pay any damages incurred. The only exceptions to this requirement would be cases involving malfeasance (unlawful acts) or willful neglect.

<u>Iowa Statutory Law</u>
Tort Liability of Governmental Subdivisions

613A.1 <u>Definitions</u>. As used in this chapter, the following terms shall have the following meanings:

1. "Municipality" means city, county, township, school district, and any other unit of local government.
2. "Governing body" means the council of a city, county board of supervisors, board of township trustees, local school board, and other boards and commissions exercising quasi-legislative, quasi-executive, and quasi-judicial power over territory comprising a municipality.
3. "Tort" means every civil wrong which results in wrongful death or injury to person or injury to property or injury to personal or property rights and includes but is not restricted to actions based upon negligence; error or omission; nuisance; breach of duty, whether statutory or other duty or denial or impairment of any right under any constitutional provision, statute or rule of law.
4. "Officer" includes but is not limited to the members of the governing body.

613A.2 <u>Liability imposed</u>. Except as otherwise provided in this chapter, every municipality is subject to liability for its torts and those of its officers, employees, and agents acting within the scope of their employment or duties, whether arising out of a governmental or proprietary function.

A tort shall be deemed to be within the scope of employment or duties if the act or omission reasonably relates to the business or affairs of the municipality and the officer, employee, or agent acted in good faith and in a manner a reasonable person would have believed to be in and not opposed to the best interests of the municipality.

For the purposes of this chapter, employee includes a person who performs services for a municipality whether or not the person is compensated for the services, unless the services are performed only as an incident to the person's attendance at a municipality function.[8]

613.17 <u>Good Samaritan law</u>. Any person, who in good faith renders emergency care or assistance without compensation at the place of an emergency or accident, shall not be liable for any civil damages for acts or omissions unless such acts or omissions constitute recklessness. (63 G.A., Chp. 292 (H.F. 39), sec. 1)

---

7. Ibid., p. 56
8. State of Iowa, School Laws of Iowa, Chapter 613A. Des Moines, Iowa, 1977.

Liability Insurance

Iowa law allows school districts to purchase liability insurance to cover them against tort suits. The policies cover all employed personnel of the district. The great majority of school districts in Iowa presently carry liability insurance. <u>Liability insurance is a protection, but it should be regarded as secondary to the teacher's conduct.</u>

  613A.7 <u>Insurance</u>. The governing body of any municipality may purchase a policy of liability insurance insuring against all or any part of liability which might be incurred by such municipality or its officers, employees and agents under the provisions of section 613A.2 and section 613A.8 and may similarly purchase insurance covering torts specified in section 613A.4. The premium costs of such insurance may be paid out of the general fund or any available fund or may be levied in excess of any tax limitation imposed by statute.[9]

EMPLOYER-EMPLOYEE RELATIONSHIP

Prior to 1967 under the *Sovereign Immunity Doctrine* public school districts could not be sued for torts committed by the district itself or by its agents or employees. The sovereign immunity doctrine stated that any governmental operation could do no wrong and therefore could not be sued without its consent.

All this was changed in 1967 when the second general session of the Iowa Legislature enacted Chapter 613.A into the Code of Iowa.

  613A.8 <u>Officers and employees defended</u>. The governing body shall defend any of its officers, employees and agents, whether elected or appointed and, except in cases of malfeasance in office, willful and unauthorized injury to persons or property, or willful or wanton neglect of duty, shall save harmless and indemnify such officers, employees, and agents against any tort claim or demand, whether groundless or otherwise, arising out of an alleged act or omission occurring within the scope of their employment or duties. Any independent or autonomous board or commission of a municipality having authority to disburse funds for a particular municipal function without approval of the governing body shall similarly defend, save harmless and indemnify its officers, employees, and agents against such tort claims or demands.[10]

  The duty to defend, save harmless, and indemnify shall apply whether or not the municipality is a party to the action and shall include but not be limited to cases arising under Title 42 United States Code Section 1983.

  613A.9 <u>Compromise and settlement</u>. The governing body of any municipality may compromise, adjust and settle tort claims against the municipality, its officers, employees and agents, for damages under section 613A.2 or 613A.8 and may appropriate money for the payment of amounts agreed upon.[11]

  280.10 <u>Eye-protective devices</u>. Every student and teacher in any public or nonpublic school shall wear industrial quality eye-protective devices at all times while participating, and while in a room or other enclosed area where others are participating, in any phase or activity of

---

9. Ibid.
10. Ibid.
11. Ibid.

a course which may subject the student or teacher to the risk or hazard of eye injury from the materials or processes used in any of the following courses:

1. Vocational or industrial arts shops or laboratories.
2. Chemical or combined chemical-physical laboratories involving caustic or explosive chemicals or hot liquids or solids when risk is involved.

Visitors to such shops and laboratories shall be furnished with and required to wear the necessary safety devices while such programs are in progress.

It shall be the duty of the teacher or other person supervising the students in said courses to see that the above requirements are complied with. Any student failing to comply with such requirements may be temporarily suspended from participation in the course and the registration of a student for the course may be canceled for willful, flagrant or repeated failure to observe the above requirements.

The board of directors of each local public school district and the authorities in charge of each nonpublic school shall provide the safety devices required herein. Such devices may be paid for from the general fund, but the board may require students and teachers to pay for the safety devices and shall make them available to students and teachers at no more than the actual cost to the district or school.

"Industrial quality eye-protective devices," as used in this section, means devices meeting American National Standards, Practice for Occupational and Educational Eye and Face Protection promulgated by the American National Standards Institute, Inc.[12]

AVOIDING NEGLIGENT ACTS

Teaching personnel must be constantly aware of their duties as viewed by the courts. No student actions should be permitted without detailed instruction and supervision. The following list of guidelines is intended to aid teachers in carrying out their duties and to minimize their chances of becoming involved in any future legal proceedings.

1. Teachers are expected to protect the health, welfare, and safety of their students.
2. Teachers must recognize that they are expected to foresee the reasonable consequences of their inactions.
3. Teachers must carefully instruct their classes and must give careful directions before allowing students to attempt independent projects.
4. All activities must be carefully planned.
5. Teachers must be careful to relate any risks inherent in a particular laboratory experiment to students prior to their engagement in that activity.
6. Teachers should create an environment in which appropriate laboratory behavior is maintained.
7. Teachers should report all hazardous conditions to supervisory personnel immediately and insist that the conditions be corrected immediately.
8. Teachers should keep adequate records covering all aspects of the laboratory operations.

---

12. Ibid.

9. The teacher's presence in the laboratory is recommended to assure adequate safety supervision.
10. Teachers should be aware of and observe local laws and regulations that relate to laboratory activities in science.

CASE STUDIES

CASES IN WHICH THE INSTRUCTOR CONDUCT WAS JUDGED TO BE QUESTIONABLE

In the following cases, all of which took place outside Iowa, the laboratory teacher was held to be guilty of negligent conduct or his or her conduct was held to be a proper question for jury consideration. If these cases had taken place in Iowa and the teacher had been found liable, the school district or insurance carrier would have had to pay damages.

In *Mastrangelo* v. *West Side Union High School* (1975), a 16-year-old high school student was seriously injured in the school chemistry laboratory when a chemical mixture exploded in his hands.[13] The student was pulverizing a mixture of charcoal, sulfur and potassium chlorate for the specified potassium nitrate. The student had received no instruction in the danger of substitution in this kind of experiment. The student filed suit alleging negligence on the part of the instructor. The court stated that it was not unreasonable to assume that the duty of a teacher of chemistry, exercising ordinary care, includes instructing the students regarding the selection, mingling, and use of ingredients with which dangerous experiments are to be accomplished rather than to merely hand them a textbook with general directions to follow the text.

In *Reagh et al.* v. *San Francisco Unified School District* (1953), a high school student brought an action against the school district for injuries he received in the spontaneous explosion of some chemical reagents.[14] The student was enrolled in the R.O.T.C. program at the school. The student had asked his chemistry teacher for instructions on making smoke bombs to be used in the R.O.T.C. maneuvers. The student asked the instructor if it would be all right to add potassium chlorate and sugar to red phosphorus to make the smoke bomb and the instructor said yes. The student put quantities of the three chemicals into the same container. The container exploded, severely injuring the student.

The court stated that the teacher had never instructed the class in the danger of combining potassium chlorate with either sugar or red phosphorus although the teacher knew that they might explode. The court held that the question of whether the school district and its employee, the teacher, were negligent in allowing the student access to the chemical reagents without proper instructions was a legitimate question for the jury.

In *Jay* v. *Walla Walla College* (1959), a college student was seriously injured when an explosion occurred as he was trying to fight a laboratory fire.[15] The student was working on an authorized experiment when he heard an explosion across the hall. He ran across the hall with a fire extinguisher and was injured when a more serious explosion occurred. The initial explosion occurred

---

13. *Mastrangelo* v. *West Side Union High School District* (1975), 42 *Pacific Reporter*, 2d Series 634.
14. *Reagh* v. *San Francisco Unified School District* (1953), 259 *Pacific Reporter*, 2d Series 43.
15. *Jay* v. *Walla Walla College* (1959), 335 *Pacific Reporter*, 2d Series 458.

in the midst of an experiment involving ethyl ether which was being conducted by two other students. There was evidence that the professor guiding the two students failed to provide the proper supervision and direction even after he had been notified of two previous minor explosions which had occurred during the course of the experiment.

The court held that the question of whether the professor had provided the proper supervision for the students under his direction was a proper question for the jury to decide and the court affirmed the verdict for the student.

In *Bush* v. *Oscoda Area Schools* (1976), a student brought suit for personal injuries against her teacher, school principal, district superintendent, and the school district itself.[16] The student was injured when a container of methanol ignited in a classroom. A mathematics classroom was being used for a physical science class due to crowded school conditions. The room contained no storage or ventilation facilities nor any of the other equipment usually associated with a science laboratory.

Open flame alcohol lamps were used in the science experiments because gas outlets were not available. Methanol was stored in bulk in an old plastic jug which was allegedly damaged and split. The jug and the lamps were kept on a counter in the rear of the room. The student alleged that some methanol had been spilled on the counter near a lighted lamp and, as she attempted to extinguish the lamp, there was an explosion and fire which ignited her clothing and resulted in severe second and third degree burns to her person.

The student claimed various acts of negligence including the following affirmative acts:

1. Leaving spilled alcohol exposed to ignition sources;
2. Failure to properly handle and store the methanol when open flame lamps would be in use proximate thereto; and
3. Keeping the methanol in a damaged container.

The plaintiff also claimed the following acts of omission:

1. Failure to warn and supervise students in handling methanol around flame;
2. Failure to train students and school personnel in the use of the fire alarm system and fire extinguishers; and
3. Failure to have the fire alarm equipment in working order.

The court held that the school district was immune from liability under the government immunity doctrine and that the superintendent was not personally negligent in any way. As to the teacher, the court held that her conduct was of such a nature as to constitute a proper question for the jury. As to the principal, the court held that he was not responsible for the acts of the teacher but, as he was responsible for curriculum and class scheduling, he should have known of the dangers inherent in using the mathematics classroom as a physical science laboratory and, consequently, his conduct was also a proper question for the jury.

In *Pittman* v. *City of Taylor* (1977), a 16-year-old high school boy began mixing chemicals in his home that had been furnished him by teachers to construct a rocket as a science project.[17] An explosion occurred and he was

---

16. *Bush* v. *Oscoda Area Schools* et al. (1976), 250 *Northwestern Reporter*, 2d Series 759.

17. *Pittman* v. *City of Taylor*, 247 *Northwestern Reporter*, 2d Series 512 (Michigan 1977).

severely injured. At that time the common-law doctrine of governmental immunity prevailed. A 1964 Governmental Immunity Act had been declared to be unconstitutional and the legislative defect was not corrected until August 1, 1970. The boy's parents filed suit against, among others, the teachers and the board of education. In the trial court, the judge dismissed for the board of education. On appeal the supreme court reversed, abrogated the governmental immunity rule for this case and for the future but not retroactively, and remanded the case. The court said the immunity rule was court made and deserved to be terminated by the courts. It quoted this language.

> The rule of governmental immunity has as a legal defense only the argument that age has lent weight to the injust whim of long-dead kings. It is hard to say why the courts of America have adhered to this relic of absolutism so long after America overthrew monarchy itself.

In *Simmons* v. *the Beauregard Parish School Board* (1975), the school board was held responsible for "actionable negligence" for a science teacher's lack of supervision over a 13-year-old student's science exhibit.[18] The student had built a volcano using a firecracker for power and visual effects and was injured.

In *Butler* v. *Louisiana State Board of Education* (1976), a student fainted, fell, and injured six teeth after voluntarily giving blood for a biology experiment conducted by another student with the approval of the professor.[19] The Court of Appeals of Louisiana held that the professor owed a duty to the volunteer student to see that proper supervision and precautions were taken and that he had breached that duty. The court further held that the evidence failed to establish any contributory negligence on the part of the volunteer student.

The seven cases described above give examples of real teaching situations and incidents. Examples of this nature are more illustrative and forceful than a list of do's and don'ts. These seven cases can be used by every teacher as a yardstick against which their own behavior can be measured.

CASES IN WHICH THE INSTRUCTOR HAS BEEN ADJUDGED NOT LIABLE

Teachers have not always been found liable in classroom and laboratory accidents. In *Moore* v. *Minor Order Conventuals* (1959), a student was adjudged guilty of contributory negligence after he was severely injured in a chemical explosion.[20] The student, along with fellow students, had received permission from the teacher to enter a laboratory to set up equipment for an experiment to be conducted later. While in the laboratory, the student attempted to make a batch of gunpowder using a formula of his own. He mixed potassium nitrate, sulfur, manganese dioxide, and phosphorus together. An explosion occurred and he was seriously injured. The student claimed negligence on the part of the school and teacher because no instructions had been given to him with respect to safety in the laboratory and no warnings as to the dangers involved in mixing chemicals.

---

18. *Simmons* v. *Beauregard Parish School Board*, 315 *South 2d Series* 883 (Louisiana App. 1975).
19. *Butler* v. *Louisiana State Board of Education*, 331 *South 2d Series* 192 (Louisiana App. 1976).
20. *Moore* v. *Minor Order Conventuals* (1959), 267 *Federal Reporter*, 2d Series 296.

The court held that the student's injuries were the result of his own imprudent acts, and no award could be granted to him.

In *Wilhelm* v. *Board of Education* (1962), two 13-year-old students were working on science projects, with the approval of the teacher, in a laboratory with the door closed.[21] After 10 minutes, the two students began to play with some chemicals in glass bottles which were on a laboratory shelf. The students knew that the chemicals were dangerous. While they were mixing and grinding the chemicals, the mixture flared up, seriously injuring the plaintiff. The court held that the plaintiff was guilty of contributory negligence as a matter of law and disallowed his claim.

In *Madden* v. *Clouser* (1971), two students had been fighting when the teacher briefly left the room, which fight was over a pencil, where one of the students let go and as another student was turning around, the pencil went into his eye, resulting in a loss of sight to the eye, the teacher was absolved of liability, since the proximate cause of the injury was an intervening and wholly unforeseen force.[22]

In *Desmarais* v. *Wachusett Regional School District* (1971), the Supreme Judicial Court of Massachusetts failed to find a teacher of chemistry negligent by reason of alleged misfeasant conduct which resulted in an eye injury to a student who was not wearing safety glasses.[23] The court said that the standard for teachers in public schools was liability only for their own acts of misfeasance in connection with ministerial matters. The court found no duty to require the wearing of safety glasses. But even assuming that the teacher was bound to require the wearing of safety glasses, the court held he could not be held guilty of misfeasance or mere inaction. The court further declined to hold that the legislature had waived governmental immunity by enacting a statute providing indemnification of teachers for expenses or damages arising out of negligence. In Iowa there is a law mandating that students wear safety glasses under such situations.

In *Maxwell* v. *Santa Fe Public Schools* (1975), action for eye injuries was sustained for a student when a glass container exploded in science class.[24] A verdict in favor of the teacher but against the Santa Fe School District and board of education was not inconsistent.

The cases discussed give an indication of how courts view certain situations. However, the law is constantly being changed, altered, or modified to deal with changing social patterns. *The current trend in negligence cases of this nature is in favor of the plaintiffs and against the defendants.*

---

21. *Wilhelm* v. *Board of Education* (1962), 189 *Northeastern Reporter*, 2d Series 503.
22. *Madden* v. *Clouser*, 277 A. 2d Series 60. (Maryland, 1971).
23. *Desmarais* v. *Wachusett Regional School District*, 276 *Northeastern Reporter*, 2d Series 691. (Massachusetts Sup. Jud. Ct. 1971).
24. *Maxwell* v. *Santa Fe Public Schools*, 534 P. 2d Series 307. (New Mexico App. 1975).

3

Eye Protection and Eye Care

EYE PROTECTION

In many cases the teacher is bewildered as to the proper eye protection that should be used in his/her laboratory. It is really an easy question to answer. All you have to understand is the American National Standards Institute standards set up for describing to the optical industry what shall be required to produce industrial quality eye protection. (See also Appendix A, pp. 107-8.)

If you read these standards (Z87.1-1979) you may find that the indepth reading will be confusing. The things you need to know are:

I. Type of eye and face protection
   A. Spectacles
      1. Style A--No side shields.
      2. Style B--Side shields (type that should be used).
   B. Cover Goggles
      1. Dust (do not use in chemistry or physics lab).
      2. Splash (chemistry).
   C. Face shields (used in combination with A or B listed above).
   D. Laser spectacles
II. How to identify the industrial quality eye protection
   A. Spectacles *shall* have the manufacturers' trademark on both lenses. The frame fronts *shall* have the manufacturers' trademark and a Z87 logo. The temples *shall* bear the manufacturers' trademark, a Z87 logo plus the overall length.
   B. The splash (chemistry) cover goggles are identified in the same manner. The lens *shall* bear the manufacturers' trademark. The frame *shall* bear the trademark plus the Z87 logo.
   C. The face shield *shall* also bear the manufacturers' trademark on the lens. The head gear *shall* bear the trademark plus the Z87 logo.
   D. Laser spectacles *shall* meet the ANSI standards Z136.1-1976. The laser protective eye wear *shall* bear a label identifying the following:
      1. Laser wavelength.
      2. Optical density.
      3. Visible light transmission.
   Check with the manufacturer of the laser equipment as to the type of eye protection needed. Also, remember that the laser spectacles *shall* be equipped with side shields.

The student may come to you and state, "Doctor has said that my glasses are safety glasses." If they do not bear a trademark and the Z87 logo as indicated in the above information they do not meet the standards and thus are not industrial quality eye protection. Street-wear glasses are impact resistant but they will not withstand the impact test required by the industrial quality lenses. Also, the lenses in the street-wear may be put into the frame of either side. Industrial quality frames allow the lenses to be put in from the front of the frame only. The back has a higher ridge to help hold the lens in place in case of a heavy impact. *Do not allow* street-wear spectacles to be used unless they are covered with the splash type cover goggles. Do not make the mistake of accepting a letter from the doctor saying that they are safety glasses.

Contact lenses *do not* provide eye protection in the industrial sense and shall not be worn in a hazardous environment without the appropriate cover eye wear. This means that splash goggles shall be worn by the person wearing contacts. Be careful and use the type of splash goggles that has plenty of ventilation for the contact wearer. This means that there shall be six metal or plastic ports, three per side, on the frame, or it shall have a ventilation method allowing enough air in to keep the contacts functioning properly.

Do not allow visitors in the laboratory without the proper industrial quality eye protection. Do not give them a pair of "visitor glasses." These do not meet the standards. The Iowa School Eye Protection Law states that visitors shall be given eye protection that meets the Z87.1-1979 standards.

It is strongly recommended that each student buy his/her own eye protection. If this is not possible and the eye protection must be shared, then the eye protectors shall be thoroughly cleaned and disinfected as required by the Z87.1-1979 standards listed below.

6.4.3 Disinfection

6.4.3.1 <u>General</u>. When a person is assigned protective equipment it is recommended that this equipment be cleaned and disinfected regularly, as herein specified. Equipment to be shared shall be cleaned and disinfected before use by another individual.

6.4.3.2 <u>Procedure</u>. Thoroughly clean all surfaces with soap or suitable detergent and warm water. Carefully rinse all traces of soap or detergent. Completely immerse the protector for 10 minutes in a solution of modified phenol, hypoclorite, quaternary ammonium compound, or other disinfection reagent, in a strength specified by the manufacturer of the safety equipment, at a room temperature of 20° C (68° F). Remove protector from solution and suspend in a clean place for air drying at room temperature, or with heated air. Do not rinse because this will remove the residual effect.

Ultraviolet disinfecting equipment may be utilized in conjunction with the preceding washing procedure, when such equipment can be demonstrated to provide comparable disinfection.

Spray-type disinfecting solutions and bactericides may be utilized when such pressurized spray solutions can be demonstrated to provide comparable disinfection with the immersion procedure outlined above.

Protectors showing need for extensive cleaning should be disassembled to the extent possible without tools, prior to the washing and disinfection procedure. Replace defective parts with new ones.

6.4.3.3 <u>Storage</u>. The dry parts or items should be placed in clean, dustproof containers to protect them.

The following statements are also found in the Z87.1-1979 ANSI standards.

A. 4.5 Protectors should be kept clean and in good repair. A scratched

Eye Protection and Eye Care 21

      or dirty lens or a loose screw can cause a problem and maybe an accident. Keep them clean and in good repair.
B. 4.8.1 Eye and face protective devices manufactured before the approval date of this standard, and meeting all requirements of the superseded ANSI Z87.1-1968 standard, shall be allowed to continue in service throughout their useful life.

    The only basic difference between the two standards is the required Z87 logo that is required in the Z87.1-1979 standards.
    The student may ask if photo grey or photographic (photochromic) lenses may be used in the lab. Paragraph 6.3.3.3.4 (Special-Purpose Lenses) of Z87.1-1979 is very specific in answering this question; the answer is NO. You will have to be careful with street-wear glasses of this type used under the cover goggles or face shield. The darkened lens may cause an accident because the student cannot see well.
    The Iowa school eye protection law is for the safety of the student. Allow no exceptions when it comes to eye protection just as you allow no chemical or physical lab experiment to be performed when the outcome will cause injury or death to the student or students.
    If one person is doing an experiment in the lab and eye protection is needed, then every person in the lab shall wear eye protection. It is strongly suggested that the school system purchase a copy of the Z87.1-1979 standard from the American National Standards Institute, 1430 Broadway, New York, NY 10018.
    The instructor shall set the example. Wear your eye protection at all times.

E Y E   C A R E

    Safety in the classroom is vitally important to all of us. Every year, tragic human suffering and economic losses could be greatly reduced if not completely avoided if only proper safety practices were implemented and followed in our industrial and science areas.
    Every administrator and teacher involved in science education must become completely familiar with the proper eye protection, and how and when it is to be worn.
    Proper protection is of utmost importance, but of equal import in the area of protection, not only to the eyes but to the student in toto, is that of efficient vision, for if the student is to perform chemistry experiments he/she must have good vision so as not to make mistakes in reading the mixture of chemicals, or in seeing the calibrations on the test tube, as an example, or misjudging the distance and reaching for a beaker of acid and knocking it over and creating a serious burn to himself/herself or others in the classroom.
    Fifty-two percent of the American public currently requires and wears corrective lenses of some form. You must be alert to the fact that Food and Drug Administration approved dress (street) eye wear does not meet the specifications under Z87.1 of the American National Standards Institute (ANSI) and cannot be worn in the science areas. They either must be replaced with industrial eye wear meeting the Z87.1 (ANSI) standards or covered with an approved safety cover goggle. See Appendix Figure B.1, p. 115.
    Contact lenses are being worn by over 14 million Americans, and with the development of advanced designs such as the soft contact lenses, the gas permeable lenses and new materials in the hard contact lenses, the popularity of

this form of visual correction is growing by leaps and bounds. Over 50 percent of these contact lens wearers are in the age bracket of 14-24 years, thus placing these persons in the high school and college area, and many will be in your science classes.

Contact lenses are generally used to correct medium to high degrees of vision error, many of which would not be correctable to 20/20 or effective vision with conventional spectacles. Examples of these conditions are kerataconus, conical cornea anisometropia and aniseikonia (which causes unequal image sizes to the two eyes, causing double vision), certain types of strabismus (cross eyes), causing double vision, and some types of amblyopia (lazy eyes) and post-surgical cataract patients (apkakics).

In past years contact lenses have not been recommended in the educational or industrial environment where there is a chance of radiation, chemical splash or flying objects; however, with the recent advancement in types of lenses and lens materials and greater knowledge, a reversal of this position has occurred. Both the FAA and FHA have reversed their position and now allow commercial pilots and truck drivers to wear contact lenses.

The 1979 ANSI Z87.1 Practice for Occupational and Educational Eye and Face Protection in Section 4.3 of General Requirements states:

> Contact lenses of themselves do not provide eye protection in the industrial sense and shall not be worn in a hazardous environment without appropriate cover safety eyewear.

Great concern has been shown in past years with regard to contact lenses and chemical splash being trapped behind a contact lens. In a report presented by the Department of Labor Occupational Safety and Health Administration in Job Safety and Health, Vol. 3, No. 7, July 1975, researchers for Eastman tested the chemical trapping theory and found splash triggers an eyelid spasm that tightens the lens to the cornea, thus sealing off the area under the lens and working as a barrier to the irritant, protecting rather than endangering the cornea. Chemical fumes and noxious vapors generally cause no serious changes to the eyes or to contact lenses. Temporary excessive tearing may occur to all those exposed, whether wearing contact lenses, glasses or no eye wear.

Students who are required to remove their contact lenses during the periods they are in science laboratories may be visually handicapped because of spectacle blur and other aforementioned conditions, thus creating a safety hazard to themselves and others; therefore contact lenses with the proper six-vent approved cover goggles are considered safe in the science environment and should not be removed.

It is the recommendation of the Eye Care Committee of the Iowa Optometric Association that (1) all students enrolling in science courses should have a complete vision examination or at minimum a visual screening performed by the school nurse to ascertain the student has adequate vision to perform the required tasks, (2) teachers should ascertain the names of all students wearing contact lenses and make certain that they are wearing the proper approved safety goggles.

Therefore, as we consider all aspects of protection of our students in the science areas, we must:

1. Determine that all students are visually competent either by requiring complete eye examinations or a written statement from the student's eye care practitioner that the student is visually competent. At the very minimum, a visual screening should be done by the school nurse on all students enrolling in science classes.

Eye Protection and Eye Care 23

2. Provide and require that the proper approved cover goggles be worn by all students at all times that any experimentation is in process within the classroom.

    Remember that you as a teacher or administrator are subject to litigation in tort liability if a student suffers an eye injury or other injury as a result of not wearing approved eye protection.
    Protection is a must and will prevent or at least significantly minimize eye injuries, but that is not sufficient, for you must be prepared to treat an ocular emergency.
    The equipment and supplies necessary to meet these ocular emergencies are as follows:

1. Approved eye wash fountain (See Appendix Figure D.1, p. 123.)
2. Q-Tips
3. Ample supply of sterile eye patches
4. Ample supply of surgical tape
5. Eye cups
6. Posted phone number of the University of Iowa Poison Control Center
7. Posted phone number for emergency transportation
8. Posted phone number of the nearest ophthalmologist or hospital

    What are the types of ocular injuries?

1. Injury to the eyelids
2. Injury to the eye
   A. Blunt or contusion
   B. Penetrating
3. Chemical burns: the only true ocular emergency

1. Injury to the eyelid, either contusion or laceration.
   A. Stop hemorrhage: gentle direct pressure with sterile eye pads and cold compress or ice.
   B. Apply sterile pressure patch.
   C. Continue cold compress or ice to reduce swelling, hemorrhage, and pain.
   D. Call ophthalmologist or hospital.
   E. Transport for medical attention with patient in supine position, head slightly elevated above heart.
2. Injury to the eye--contusion.
   A. Do not irrigate.
   B. Apply dry sterile dressing; do not use pressure patch.
   C. Apply ice or cold compress to reduce hemorrhage.
   D. The structure of the eye may be torn or ruptured. You may wish to cover the eye to reduce eye movement.
   E. Transport as soon as possible to hospital with the patient supine, head slightly elevated.
3. Penetrating injury to the eye.
   A. Do not irrigate.
   B. Do not attempt to remove object.
   C. Cover both eyes with loose sterile eye patches to reduce movement and infection. *Do not pressure bandage.*
   D. Apply cold compress to reduce hemorrhage.
   E. Keep patient quiet.
   F. Phone ophthalmologist or hospital.
   G. Transport to ophthalmologist or hospital immediately with patient supine, head slightly elevated above heart.

4. Chemical burn to the eye--only true emergency.
   Acids cause serious damage immediately. Alkalines (bases) are the most serious, with a slower reaction time. It is sometimes hours to days before complete reaction occurs.
   A. Invert lid if possible over Q-Tip.
   B. Profuse irrigation for 10-15 minutes in approved eye wash fountain.
   C. While irrigating, have someone else call the University of Iowa Poison Control Center. Tell them what chemical entered the eye. They will inform you of probable severity of the injury and the necessary antidote.
   D. Call ophthalmologist or hospital.
   E. Transport immediately to ophthalmologist or hospital in supine position.
   F. Continue diluting the chemical by use of eye cup.

REMEMBER, chemical burns to the eyes are the only true emergency. Prompt action is necessary--IRRIGATE, IRRIGATE, IRRIGATE!

Eighty percent of our knowledge is gained through that precious gift called sight, so use only approved eye protection and not just street eye wear. Do not mandate discontinuance of contact lenses during science laboratories. You may actually be creating a visual impairment that can cause the student to have an accident. If you have any questions regarding the visual efficiency of your contact lens wearing student, contact his doctor and let him decide. Require all students to wear approved six-vent cover goggles and if an ocular injury or emergency does occur, be calm and be prepared to render the proper first aid.

In this manner you will assure better science through safety.

REFERENCES

ANSI Z87.1, *Practice for Occupational and Eye-Face Protection.*
Code of Iowa, 280.20, Eye Protection Devices for Students and Teachers.
Contact Lens News Backgrounder, American Optometric Association, May, 1979.
Occupational Safety and Health Administration, *Job Safety and Health* 3, No. 7 (July, 1975).

4

Safety in Biology Settings

GENERAL SAFETY FOR BIOLOGY TEACHER AND STUDENTS

The general principles that apply to the concern for safety in chemistry, physics, and other science classrooms and laboratories are appropriate also in the biology program.

The total responsibility for safety in the classroom and laboratory must be shared by teacher and student. It becomes a special responsibility of the teacher to develop an acceptance on the part of the students that they are expected to share actively in the maintenance of safety in their biology activities. The teacher, with greater background and more experience, will be looked to for specific information as to how to deal with existent hazards, for the anticipation of potential hazards, and for careful supervision of activities when potential hazards are expected. Also, the teacher must incorporate safety knowledges and attitudes within the objectives of the course.

There is no simple recipe for all to follow in achieving student acceptance of this shared responsibility. Both positive and negative reinforcement will quite likely be necessary. Attention to safety practices and conditions must always be serious business. Death and physical handicaps may be the result of the lack of such attention. Carelessness and an improper attitude are common reasons for accidents. The biology classroom and laboratory are places for serious work and study. They are not places for physical play.

Special attention must be given on a regularly scheduled basis to safety procedures to be followed in all classroom and laboratory activities. Regular drills should be included in the instructional program which will emphasize steps to be followed in the case of emergencies due to accidents. Examples of such drills are ones in which the following conditions in biology laboratory or field work might be involved: fire, severe burns, presence of chemical reagents in the air, hemorrhaging cuts, animal bites, and swallowing of toxic material. There are others that would be appropriate for the activities in various programs.

Unnecessary hazards are sometimes created from overcrowding in the biology laboratory. The facilities section makes reference to this. The goal to be sought is to have no more than 28 students per laboratory with each student having at least 30 square feet of working space. In addition, auxiliary space should be provided for animals, plants, reagents, and storage. Traffic movement through the laboratory and between it and the various units of auxiliary space must also be planned carefully and enforced.

Prominently displayed in the classroom, laboratory, storage areas, greenhouses, animal rooms, and other units of the biology facilities should be posters listing locations and telephone numbers of individuals and agencies to be contacted in cases of emergencies. These posters should include information concerning physicians, ambulance services, rescue squads, emergency rooms at health centers and hospitals, poison control centers, fire stations, and police stations.

Reasonable housekeeping standards of orderliness and cleanliness must be met in the care of the biology facilities. Certain areas should receive special attention. These include animal rooms, refrigerators used to store microorganisms and culture media, dishwashing and dish drying areas, and areas where cages, terraria and aquaria are washed and stored. Cleanliness and orderliness contribute not only to reduction of hazards but also to positive attitudes about safety.

BIOLOGY FACILITIES

Many of the recommendations concerning the reduction of hazards in biology facilities are included in the general safety rules described above and in the section devoted to all science facilities. Reference to these parts should be made in considering this topic.

Storage of materials is an important concern in biology, as in the other sciences, and there are a number of special precautions to be observed. Reference should be made to the section on chemical hazards and the precautions to follow in storage of all chemicals. There are a number of volatile liquids, such as alcohols and ethers, regularly used in biology for which special designated space should be provided. Special attention should be given to ventilation and exhausting of gases in this space. Drugs should be used sparingly, if at all, in elementary biology and physiology laboratory work. If any are to be stored, special locked space should be provided and a careful inventory maintained and kept in a secure place in the instructor's or department's office.

Lockable refrigerators are important storage spaces in biology. Most facilities should have more than one, with clear designation of the kind of storage for which each is to be used. Areas within refrigerators, e.g., shelves and compartments, should be plainly marked as to what is to be stored there and some impressive label should be used to designate materials which are prohibited in these areas. Foods should not be permitted in refrigerators where microorganisms or chemicals are being kept. In general, foods for human consumption should not be kept in biology refrigerators regardless of what kinds of materials are being stored or held there.

Preserved specimens, both animals and plants, should be stored where they can be regularly inspected. Ventilation and exhaust systems should be provided and there should be ready access for inspection and for additions and withdrawals. Access by students should be under the control of the teacher. Specimens being used during several sessions should be stored in bins and/or containers with corrosion-proof liners and should have preservative material in them.

As was indicated in the introductory section on general principles, special areas should be provided for the holding of live animals. If special provisions are not available, animals should not be kept in the classroom or laboratory. A special area does not necessarily mean a completely separate room. It should be a definitely designated and localized area, divided in some fashion from the rest of the facility. Necessary space for cleaning cages, "grooming" animals, and preparing food should be available. Tamper-proof locks should be used on most cages. Although an increasing number of secondary school biology facilities have some kind of greenhouse, the segregation of plants from the classroom is not necessary to the same degree as it is with animals. The value of the greenhouse, of course, is for culture or

Safety in Biology Settings

growing of plants throughout the year. In the holding of both plants and animals in or near the classroom, the possibility of some students having allergic reactions to the presence of specific organisms or a large number of kinds of them must be considered.

Bulk storage areas, especially, should be kept locked and carefully monitored with inventorying being done regularly and frequently. Again, this inventory should be kept in a secure place in the instructor's or the department's office.

Labeling is an important factor in avoiding hazardous use of laboratory materials. All things kept in closed containers, e.g., chemicals, preserved specimens, microorganisms, should be clearly labeled. If unlabeled materials cannot be unequivocally identified, they should be discarded, using proper disposal methods.

Areas of the laboratory to be used for specific purposes should be marked with appropriate signs. Instruments or appliances to be used for specific purposes should also be marked with appropriate signs. These signs should be large enough to be noticeable and the lettering should be clearly legible.

The disposal of used or unusable materials from a biological laboratory requires careful attention. Special covered containers should be provided for discarded biological material, for example, remains from dissection studies. These materials should be burned or buried. Some biological facilities will have special incinerators. Such things should not be burned in a trash burner in the open air. If they cannot be burned, they should be buried and the depth of burial should be at least three feet.

Chemicals associated with preserved materials should be washed down a drain with large amounts of water. Other chemicals should be disposed of in accordance with the recommended procedures listed in such publications as *Laboratory Waste Disposal Manual*, Manufacturing Chemists Association, 1825 Connecticut Avenue, N.W., Washington, D.C. 20009. The section dealing with chemical hazards will have other procedures and references of value in disposing of chemicals associated with biological activities.

Disposable petri dishes should be incinerated if at all possible. Glassware containing microorganisms (such as glass petri dishes) should be sterilized before washing. Steam should be used with 15 pounds per square inch guage pressure for at least 15 minutes. More details for this will be provided in the special section on microorganisms.

Cleanliness is important in all science classrooms and laboratories, but it is especially important in biology where living material is being used. Cleanliness is necessary to prevent deleterious effects upon students from coming in contact with plants or animals that are diseased, or carriers of disease-producing agents, due to nonhygienic conditions. The impact of non-esthetic conditions, such as odors or unsightly offal, in addition to being a health hazard, may have exaggerated effects upon attitudes about biology and working with living things. Also, cleanliness is necessary in order for plants and animals to thrive. They, too, may become diseased and suffer from the absence of hygienic conditions.

Animal cages present some special problems in being kept clean. Some attention will again be given to this in a later section dealing with animals in the laboratory. The daily cleaning of cages in use must remove excreta and unusable food and other waste material. This must be disposed of, if possible, by burning or burying. If garbage is regularly collected from the biology rooms and disposed of through an approved standard garbage disposal system, this method could be used if there is no disease or other unusual condition involved. Cages should be frequently washed through immersion in cleansing solution and rinsed thoroughly. Since cages need to be washed regularly with water and other liquid cleansing materials, it is best if they are made of stainless steel, aluminum, or other rustproof material. When cages are taken out of regular use they should be carefully cleaned. Never store cages without cleaning them. Again, before they are put back in use, they should be cleaned and disinfected.

Aquaria and glass terraria must be carefully maintained. A number of works listed in the references describe the proper care and maintenance of aquaria and terraria when they are holding animals. Special attention should be given to storage of them when they are not being actively used. They become safety hazards if they are placed where they can fall and cut someone or where they can be stepped on or into and cut someone in this manner. As with cages, these materials should always be cleaned before they are stored. Mineral deposits may usually be removed by a weak acid solution, such as vinegar or dilute hydrochloric acid. Careful rinsing should follow the dissolution of these materials. Organic debris left in them is more of a potential health hazard to both people and the next organisms that will inhabit them because it becomes a potential medium for the growth of pathogenic microorganisms.

Some attention has already been given to the provision of certain areas for specific purposes, such as areas for holding live animals, areas for keeping preserved specimens, areas for bulk storage, and other such uses. In addition, when special activities are being worked with, special areas may need to be reserved. Work with microorganisms could be an example of this. Work with physiological responses of organisms to certain chemicals might be another. Of particular concern insofar as the avoidance of a hazard is concerned is to have an activity that might possibly have some potential of hazard to the student be well-defined in space utilization so that it may be more carefully monitored. The activity should be contained in a space that can be cleaned when the activity is over.

In the discussion of general safety rules, traffic control was mentioned. This is associated with the limitation in the numbers of students in special activity areas. Desirable activities may be worked with in somewhat limited space, but monitoring and careful supervision may be necessary in order to avoid hazards.

MICROORGANISMS

Known pathogenic organisms should not be used. All cultures, however, should be treated as though they might contain pathogens. Nonpathogenic cultures may become accidentally contaminated. A cough or a sneeze while working with them may be enough for such contamination. Proper sterile techniques should always be used. When working with microorganisms a *properly* working exhaust hood and/or *properly* worn surgical mask should be used.

Petri dishes or other containers to be used for culturing should be carefully disinfected with such a chemical as phenol before being washed. Laboratory glassware and equipment known to be contaminated with hazardous material, or suspected of being so, should be autoclaved first, then cleaned. Disinfectants are sometimes ineffective if the contamination level is high or organic matter is present. The washing should use a household detergent. Following this, sterilization in an autoclave should be done. In the absence of an autoclave, a pressure cooker can be used. Care must be exercised in the use of either an autoclave or any steam pressure sterilizer or cooker. The pressure relief valve should be checked, and maximum pressure should be limited to no more than 20 pounds per square inch. The dishes can be adequately sterilized by using steam pressure at 15 pounds for 15 minutes. It should be remembered that pressure does not sterilize. If, for example, the flow of live steam in an autoclave is blocked by debris in the drain, the pressure will rise to the set limit and remain there for the time of the cycle, but the temperature will probably not reach sterilizing levels.

The media used should also be sterilized and carefully transferred to the culture dishes. Wire loops used in transferring microorganisms should be flamed before and after use. Open flames should be used sparingly and carefully in laboratories, but transfer loops may be sterilized by such open flames if only nonpathogens are involved. Gas flames are better than alcohol and especially better than candle flames, since candles deposit carbon on the loop. (See Appendix D, pp. 129-30.)

It should be noted that aerosols are formed when loops are sterilized in open flames. If virulent organisms have been transferred with the loops, the suspended material may contaminate the air. Even if the use of pathogens is avoided, the possibility of such contamination should be considered.

If cultures are to be made available for observation and especially if they are to be passed around in a class, the petri dishes or other containers should be sealed tight with transparent tape.

After a particular activity using microorganisms is completed, the culture should be sterilized with steam pressure, the containers should be disinfected, sterilized, and washed before storage. Rubber gloves should be worn while this is being done. The laboratory area where the work was done should be carefully cleaned and disinfected. Students should also wash carefully before leaving the area. Students working with microorganisms in the laboratory should wear surgical masks.

FIELD ACTIVITIES

Reference should be made to the section emphasizing safety in the field. More accidents happen while studying biology in the field than in the laboratory. Careful preparation for field work and special assignments to individual students for attention to safety measures will help reduce accidents in the field.

A greater variety of safety hazards exist in the field than in the laboratory. Plants, animals, weather conditions, topographic features, and many other factors need to be considered in planning field activities and in the actual field work. This section, however, will emphasize plants and animals in the laboratory, but many of the precautions and suggestions would apply also to field conditions.

POISONOUS PLANTS

Again, attention should be given to hazards described in the field trip section and precautions to follow.

Some plants may be considered poisonous through contact. Most commonly thought of in this category is poison ivy. Students should be carefully instructed as to its characteristics in order that it can be readily recognized in the field. The best safety rule in preventing poison ivy irritation and infection is to avoid contact with it. Although not all students will be sensitive to it, field activities should be conducted with the premise that all will be irritated by it. The use of protective clothing, particularly gloves, should be encouraged.

If it is known that contact has been made with poison ivy, thorough washing with strong soap or detergent solutions should be done as soon as possible. A number of soothing lotions may be obtained from local drug stores or physicians.

HOUSE PLANTS

Plant	Toxic Part	Symptoms and Comment
Castor bean	Seeds	Burning sensation in mouth and throat. Two to four beans may cause death. Eight usually lethal. Death has occurred in U.S.
Dieffenbachia (dumbcane), caladium, elephant's ear, some philodendrons	All parts	Intense burning and irritation of mouth, tongue, lips. Death from dieffenbachia has occurred when tissues at back of tongue swelled and blocked air passage to throat. Other plants have similar but less toxic characteristics.
Mistletoe	Berries	Can cause acute stomach and intestinal irritation. Cattle have been killed by eating wild mistletoe. People have died from "tea" of berries.
Poinsetta	Leaves, flower	Can be irritating to mouth and stomach, sometimes causing vomiting and nausea, but usually produces no ill effects.

VEGETABLE GARDEN PLANTS

Plant	Toxic Part	Symptoms and Comment
Potato	Vines, sprouts (green parts), spoiled tubers	Death has occurred from eating large amounts of green parts. To prevent poisoning from sunburned tubers, green spots should be removed before cooking. Discard spoiled potatoes.
Rhubarb	Leaf blade	Several deaths from eating raw or cooked leaves. Abdominal pains, vomiting and convulsions a few hours after ingestion. Without treatment, death or permanent kidney damage may occur.

ORNAMENTAL PLANTS

Plant	Toxic Part	Symptoms and Comment
Atropa belladonna	All parts, especially black berries	Fever, rapid heartbeat, dilation of pupils, skin flushed, hot and dry. Three berries were fatal to one child.
Carolina jessamine, yellow jessamine	Flowers, leaves	Poisoned children who sucked nectar from flowers. May cause depression followed by death through respiratory failure. Honey from nectar also thought to have caused three deaths.
Daphne	Berries (commonly red, but other colors in various species), bark	A few berries can cause burning or ulceration in digestive tract causing vomiting and diarrhea. Death can result. This plant considered "really dangerous," particularly for children.
English ivy	Berries, leaves	Excitement, difficult breathing and eventually coma. Although no cases reported in U.S., European children have been poisoned.
Golden chain (laburnum)	Seeds, pods, flowers	Excitement, intestinal irritation, severe nausea with convulsions and coma if large quantities are eaten. One or two pods have caused illness in children in Europe.
Health family (some laurels, rhododendron, azaleas)	All parts	Causes salivation, nausea, vomiting and depression. "Tea" made from two ounces of leaves produced human poisoning. More than a small amount can cause death. Delaware Indians used wild laurel for suicide.

Plant	Toxic Part	Symptoms and Comment
Holly	Berries	No cases reported in North America, but thought that large quantities may cause digestive upset.
Jerusalem cherry	Unripe fruit, leaves, flowers	No cases reported, but thought to cause vomiting and diarrhea. However, when cooked, some species used for jellies and preserves.
Lantana	Unripe greenish-blue or black berries	Can be lethal to children through muscular weakness and circulatory collapse. Less severe cases experience gastrointestinal irritation.
Oleander	Leaves, branches, nectar of flowers	Extremely poisonous. Affects heart and digestive system. Has caused death even from meat roasted on its branches. A few leaves can kill a human being.
Wisteria	Seeds, pods	Pods look like pea pods. One or two seeds may cause mild to severe gastrointestinal disturbances requiring hospitalization. No fatalities recorded. Flowers may be dipped in batter and fried.
Yew	Needles, bark, seeds	Ingestion of English or Japanese yew foliage may cause sudden death as alkaloid weakens and eventually stops heart. If less is eaten, may be trembling and difficulty in breathing. Red pulpy berry is little toxic, if at all, but same may not be true of small black seeds in it.

TREES AND SHRUBS

Plant	Toxic Part	Symptoms and Comment
Black locust	Bark, foliage, young twigs, seeds	Digestive upset has occurred from ingestion of the soft bark. Seeds may also be toxic to children. Flowers may be fried as fritters.
Buckeye, horsechestnut	Sprouts, nuts	Digestive upset and nervous symptoms (confusion, etc.). Have killed children but because of unpleasant taste are not usually consumed in quantity necessary to produce symptoms.
Chinaberry tree	Berries	Nausea, vomiting, excitement or depression, symptoms of suffocation if eaten in quantity. Loss of life to children has been reported.
Elderberry	Roots, stems	Children have been poisoned by eating roots or using pithy stems as blowguns. Berries are least toxic part but may cause nausea if too many are eaten raw. Proper cooking destroys toxic principle.
Jatropha (purge nut, curcas bean, peregrina, psychic nut)	Seeds, oil	Nausea, violent vomiting, abdominal pain. Three seeds caused severe symptoms in one person. However, in others as many as 50 have resulted in relatively mild symptoms.
Oaks	All parts	Eating large quantities of any raw part, including acorns, may cause slow damage to kidneys. However, a few acorns probably have little effect. Tannin may be removed by boiling or roasting, making edible.

Plant	Toxic Part	Symptoms and Comment
Wild black cherry, chokecherries	Leaves, pits	Poisoning and death have occurred in children who ate large amounts of berries without removing stones. Pits or seeds, foliage and bark contain HCN (prussic acid or cyanide). Others to beware of: several wild and cultivated cherries, peach, apricot and some almonds. But pits and leaves usually not eaten in enough quantity to do serious harm.
Yellow oleander (be-still tree)	All parts, especially kernels of the fruit	In Oahu, Hawaii, still rated as most frequent source of serious or lethal poisoning in man. One or two fruits may be fatal. Symptoms similar to fatal digitalis poisoning.

FLOWER GARDEN PLANTS

Plant	Toxic Part	Symptoms and Comment
Aconite, monkshood	Roots, flowers, leaves	Restlessness, salivation, nausea, vomiting, vertigo. Although people have died after eating small amounts of garden aconite, poisoning from it is not common.
Autumn crocus	All parts, especially bulbs	Burning pain in mouth, gastrointestinal irritation. Children have been poisoned by eating flowers.
Dutchman's breeches (bleeding heart)	Foliage, roots	No human poisonings or deaths, but a record of toxicity for livestock is warning that garden species may be dangerous.
Foxglove	All parts, especially leaves flowers, seeds	One of the sources of the drug digitalis. May cause dangerously irregular heartbeat, digestive upset and mental confusion. Convulsions and death are possible.
Larkspur, delphinium	Seeds, young plant	Livestock losses are second only to locoweed in western U.S. Therefore, garden larkspur should at least be held suspect.
Lily-of-the-valley	Leaves, flowers, fruit (red berries)	Produces glycoside like digitalis, used in medicine to strengthen the beat of a weakened heart. In moderate amounts, can cause irregular heartbeat, digestive upset and mental confusion.
Nicotiana, wild and cultivated	Leaves	Nervous and gastric symptoms. Poisonous or lethal amounts can be obtained from ingestion of cured smoking or chewing tobacco, from foliage of field-grown tobacco or from foliage of garden variety (flowering tobacco or nicotiana).

WILD PLANTS

Plant	Toxic Part	Symptoms and Comment
Baneberry (doll's-eyes)	Red or white berries, roots, foliage	Acute stomach cramps, headache, vomiting, dizziness, delirium. Although no loss of life in U.S., European children have died after ingesting berries.
Death camas	Bulbs	Depression, digestive upset, abdominal pain, vomiting, diarrhea. American Indians and early settlers were killed when they mistook it for edible bulbs. Occasional cases still occur. One case of poisoning from flower reported.

Plant	Toxic Part	Symptoms and Comment
Jack-in-the pulpit, skunk cabbage	All parts, especially roots	Contains small needle-like crystals of calcium oxalate and causes burning and severe irritation of mouth and tongue.
Jimsonweed (thornapple)	All parts, especially seeds and leaves	Thirst, hyper-irritability of nervous system, disturbed vision, delirium. Four to five grams of crude leaf or seed approximates fatal dose for a child. Poisonings have occurred from sucking nectar from tube of flower or eating fruits containing poisonous seeds.
Mayapple (mandrake)	Roots, foliage, unripe fruit	Large doses may cause gastroenteritis and vomiting. Ripe fruit is least toxic part and has been eaten by children—occasionally catharsis results. Cooked mayapples can be made into marmalade.
Nightshades, European bittersweet, horse nettle (solanum)	All parts, especially unripe berry	Children have been poisoned by ingesting a moderate amount of unripe berries. Digestive upset, stupefication and loss of sensation. Death due to paralysis can occur. Ripe berries, however, are much less toxic.
Poison hemlock	Root, foliage, seeds	Root resembles wild carrot. Seeds have been mistaken for anise. Causes gradual weakening of muscular power and death from paralysis of lungs. Caused Socrates' death.
Pokeweed (pigeonberry)	Roots, berries, foliage	Burning sensation in mouth and throat, digestive upset and cramps. Produces abnormalities in the blood when eaten raw.
Water hemlock (cowbane, snakeroot)	Roots, young foliage	Salivation, tremors, delirium, violent convulsions. One mouthful of root may kill a man. Many persons, especially children, have died in U.S. after eating this plant. Roots are mistaken for wild parsnip or artichoke.

Fig. 4.1. Common Poisonous Plants. (Reprinted by permission of National Safety Council, *Family Safety* 48, no. 1 (Spring 1979): 18-19.)

A number of other plants are similar to poison ivy in the numbers of people they will affect. These include poison sumac and poison oak. Irritations from these should be treated the same as the ivy.

Many other plants may be irritating to certain people upon contact. All of these irritants are usually encountered only in field work. It would be unwise in most situations to grow these plants in the laboratory or to bring uncovered samples into the classroom. Carefully prepared and sealed herbarium mounts might be used to familiarize students with them in order that they can learn to avoid them. If a site is available on the school grounds or nearby where plants might be grown for demonstration, this, like the herbarium mount, might prove valuable in instruction. Such a demonstration plot should be prominently marked.

Some plants may be considered poisonous through ingestion. Students should never eat any part of a plant being grown in the biology laboratory or greenhouse. This includes garden plants since experiments may be in progress with them that would make them toxic, or, at least, irritating. In spite of the emphasized prohibition, if a student should swallow some part of a plant, contact the poison control center for instruction. Only carelessness on the part of the teacher and student would create such an emergency as this in a biology classroom. Units of study might be included in the biology curriculum in which toxic parts and products of plants could be studied. A number of references are available that could be used in such a study. One is *Deadly Harvest: A Guide to Common Poisonous Plants* by John M. Kinsbury. Figure 4.1 is a list of common poisonous plants with diagrams of the plants and descriptions of toxic parts and symptoms of the poisoning.

Some students may be allergic to pollens, seeds, or other parts of plants being kept in the classroom. These plants should be removed when the allergic response is detected. Molds and other types of plants that flower may be responsible for individual allergic reactions.

ANIMALS IN THE LABORATORY

Poisonous animals, such as rattlesnakes, should never be kept in the elementary or secondary biology laboratory. Diseased animals should never be kept their either.

As has been discussed in the section dealing with biology facilities, special areas in the biology suite should be reserved for holding animals. Reference has also been made to the necessity of keeping the area, the cages, and the animals clean. Cages should be made of material which does not permit sticking fingers through the meshwork to attract the attention of the animals. Prominent signs should be posted in the area which express prohibitions of sticking fingers into cages and feeding the animals.

Fastidious care in feeding, watering, separating new litters, cleaning, and other such responsibilities of husbandry should be given laboratory animals. Special attention should be given to providing care during weekends and other times when school is not in session. Leather gloves should always be used in handling live animals. This might seem unnecessary with small pet-like animals, but severe infections may result from the bite of an animal such as a small mouse or guinea pig. One of the most common incidents requiring first aid attention in the science laboratory is animal bites. It is important that animals always be handled gently.

In the event an animal does bite someone, the wound, no matter how slight, should be treated by the school nurse and an accident report should be filled out. The animal should be carefully observed over a period of several days.

Special caution should be used in bringing feral animals into the classroom and keeping them there. They often do not respond well to caging and are difficult to handle, and care of them is more complicated. Most animals to be used in laboratory activities or kept for observation should be secured from a reputable pet dealer or a recognized breeder and supplier of animals for scientific purposes.

Experiments or activities with animals should be designed carefully and monitored thoroughly by the teacher to assure that they are not mistreated. Guidelines of the Animal Welfare Institute and local animal protection groups should be followed. An excellent reference is *Guide to the Care and Use of Laboratory Animals*, Department of Health, Education and Welfare, Publication No. (NIH) 78-23, Revised 1978. This is of significance not only to guarantee humane treatment of animals, it also reduces the possibility of aggressive reactions of animals to cruel and painful treatment that may result in injury to both animal and student. An injured animal in pain is often a dangerous animal.

Also, diseased animals often react violently. Some diseases may affect students, and particularly may produce infections through bites and scratches.

If animals are kept and used in biology activities, a veterinarian should be contacted and arrangements made for regular checking and for treatment of disease conditions and for help in the case of emergencies.

It often becomes necessary to dispose of animals that have been involved in experiments. Most local humane societies have facilities for this kind of "sacrifice" and disposition. If this kind of help is not available, the use of excessive doses of an anesthetic, such as methyl chloroform, is usually an accepted method of killing. The carcasses should then be burned or buried. Only in rare instances should these cadavers be dissected. Sometimes the experiment involves examination of internal tissues or organs. Extreme caution should be used in such dissection, with rubber gloves always used and the instruments carefully disinfected and sterilized after they are used.

Animals that die should be removed promptly from cages. Always use gloves. If possible, have a veterinarian check the dead animals and the animals that have been in contact with them. The veterinarian may have facilities for disposal of the dead. Incineration or burial can be used.

In addition to the usual physical and chemical hazards found in the biology laboratory, special forms of biohazards may be encountered. There are approximately 200 diseases of animals that are transmissible to man. Some the more common ones are included in a report of the 1979 AIHA Biohazard Committee prepared by its chairperson, Professor J. R. Songer, Iowa State University.

Many human diseases also infect animals and the animals then become sources of infection for other humans. For example, supposedly normal monkeys at Iowa State University have been found infected with human infections.

Lymphocytic choriomeningitis (LCM), a viral disease of mice, has occurred in humans when exposed to reportedly normal hamsters. A classical outbreak occurred several years ago when normal hamsters were kept in the Xerox room of a laboratory. Personnel waiting to use the Xerox machine fraternized with the hamsters. Before the source was discovered, several people were infected with LCM. Infection is often not apparent in rodents.

Appropriate practice should be to (1) obtain animals from certified disease-free sources, (2) become familiar with all zoonotic and human diseases for which the animal can serve as host, (3) keep animals isolated from students, if possible, and (4) keep animals used in the laboratory isolated from wild animals, which serve as carriers of many diseases.

Mycoses and mycotoxicoses can occur when handling plant and animal matter contaminated with fungal spores. There are approximately 50 species of fungi which are agents of infectious disease in man and many more are toxigenic. One must be cautious when collecting and growing microorganisms from the environment. For example, *Histoplasma capulatum*, which can be isolated from soil, can cause serious disease in humans.

In addition to infectious diseases, allergic disease can result from exposure to a number of biological substances. Many different mammals have been implicated in allergic disease including: cats, dogs, horses, cattle, sheep, goats, pigs, rabbits, rats, mice, hamsters, gerbils, guinea pigs, cheetahs, and chinchillas.

Allergic diseases and intoxication can occur by exposure to invertebrate animals also. Ticks, spiders, scorpions, bees, and wasps account for many illnesses each year.

KILLING JARS FOR STUDY OF INSECTS

Collection of insects is a popular activity of science curricula from elementary school through graduate study in entomology. For proper mounting and preservation it is important to use effective killing agents. Although advanced entomology students will use cyanide compounds, these should never be used in elementary and secondary schools. Carbon tetrachloride, once a popular killing agent for insects, should never be used.

A good killing jar for elementary and secondary school students should have a wide mouth and a secure, screw-type cover. A used mayonnaise jar wrapped with adhesive tape works very well. For most purposes, several layers of facial tissue can be put in the bottom of the jar. To this can be added six to eight drops of ethyl alcohol, formaldehyde, or ether. Absorbent type of paper, such as filter paper, may be placed over the moistened tissue paper to prevent insects from coming directly in contact with the chemical which might get them damp and might cause colors to be affected and drying to be hindered. The longevity of the killing effectiveness of this type of jar is limited. It can easily be recharged, of course. Insects should be killed quickly enough that they do not damage their wings or other anatomical structures and make the specimen a worthless addition to a collection. The top of the jar should be labeled to show that it is to be used for killing insects and the killing agent used should be included on the label.

A similar jar with a longer life of effectiveness can be made by pouring one or two inches of plaster of paris into the bottom of the jar. As the plaster hardens, the excess moisture can be absorbed or poured off. After it has hardened, it can be saturated with ethyl acetate. Again, a dry piece of paper can be placed over the plaster to keep the insects dry. This will last considerably longer than the jars using moistened paper. A label, as usual, should be placed on the lid. The fumes from the acetate are a little more pungent, but are not harmful unless breathed intensely.

Killing jars are made essentially to be used in the field. Even though rather benign substances are suggested here, killing jars should not be used in interior spaces unless these spaces are well ventilated. The vapors from chemicals such as these should not be inhaled deeply. Most of the jars charged as described here present little or no problem in disposal. Careful disposal of the paper or plaster material and thorough washing of the jar with detergent is usually satisfactory. It is important not to store them without proper disposal of materials inside them and cleansing. Of course, special methods involving dilution and burial should be used with cyanide bottles. It should again be emphasized, however, that cyanide should not be used as a killing agent in elementary and secondary school activities.

In killing stinging insects, very careful techniques should be employed. Classes should be surveyed to learn whether any students are especially sensitive to insect stings or insect "bites." If some are, explicit directions should be obtained from the family and family physician as to how they are to be given first aid and treatment. All stings should receive first aid treatment and the student carefully watched. Certain insects, ticks, spiders, and mites are possible vectors for human diseases and should be collected only with the greatest care.

PRESERVED PLANT AND ANIMAL MATERIALS

Most preserved materials for use in biology activities can be secured from a regular laboratory supply firm. Be sure to note the composition of the preserving solution. If it is not labeled, labeling information should be obtained from the firm. Specimens may be preserved in the field or in the laboratory. Most animals other than mammals may be placed directly in large jars containing preserving solution. The solution is usually prepared by diluting formalin. Formalin is a 40% solution of formaldehyde gas in water. A 10% solution of formalin is a common strength for use as a preserving and fixing

Safety in Biology Settings

agent. Vapors from formalin or formalin solutions should not be breathed directly because they can be very irritating to the nasal and throat membranes. The section on chemicals outlines details of dangerous exposure levels. Special care should be exercised in placing animals in the solution, since there may be some splashing of liquid out of the containers. As some individuals may be sensitive to formalin-based solutions, avoid extensive skin contact. If formalin solution gets in the eyes, it should be washed out thoroughly and immediately. This washing should be for more than 15 minutes and an eyewash fountain should be used, if possible. Labels should be placed at once on the containers. The labels should identify specimens and sites of collection and identify chemicals involved in preservation and fixation. Several sourcebooks are available to give details of preparation and use. The emphasis made here is that all these chemicals are to be respected and used with great care. The same sort of individual protection should be expected with these as with chemical activities, even though some of these uses are made in the field and not in the laboratory.

Many preserved specimens become a part of a rather permanent collection or museum for later study and reference. Such a collection deserves a special storage area and access to it should be controlled. Other specimens are secured for intensive study through dissection.

Special precautions are necessary in dissection. First, the specimens should be thoroughly washed before use. Some suggest they should be flushed with running water for at least 24 hours. This is probably longer than necessary, but they should be washed and flushed long enough to prevent irritation of eyes, and nasal and throat membranes while dissecting.

Second, the dissecting instruments should be in good condition. Incision accidents are less likely to happen if scalpel and scissors are sharp than if they are dull.

Third, rules of student behavior during dissection are to be strictly enforced. These should include the following: rubber gloves are to be worn; always cut with the scalpel away from you; never "play" with dissecting instruments (dissecting needles can produce dangerous puncture wounds); carefully prepare the specimen for storage if it is to be used again.

Special containers should be available to store specimens which will be used again, and weak solutions of preserving agents should be in these containers. Refrigerators may be used for storage areas, if there is no danger of contamination of other materials.

Whenever preserved materials are to be "thrown away," the solutions should be thoroughly flushed down the drain and the specimens should be incinerated or buried. Here, as with the disposal of other biological materials, the school's regular trash disposal system may be used, but it is the biology teacher's obligation to be familiar with the adequacy of protection that such a system provides.

Study skins are sometimes prepared from collected mammals. A professional mammalogist will often use an arsenic compound powder on the skin to protect it from insect damage. Secondary school collections should not use such a toxic material, but should depend on moth balls or Borax. Borax is an effective drying agent and preservative for elementary student use.

Although animals killed on roads and highways are often the source of valuable study material for mammalogists, parasitologists, and other professional biologists, it would be best to avoid their use in secondary school classes. If individual students work with them, they should be carefully instructed in precautions to use to avoid infection and exposure to virulent organisms hosted by the dead and decomposing animal.

Most attention in this section has been directed toward animals. Plant specimens are often preserved in dried form. Other than allergic reactions, there are few hazards in the preparation and use of herbarium specimens. Some plants, of course, or parts of plants, are preserved and fixed in liquid solutions. Similar precautions would be followed with these as with the animals.

SPECIFIC LABORATORY TECHNIQUES

Pipetting

Although a number of harmless substances may be pipetted, students should never pipette by mouth. All pipetting should use an aspirator bulb or pipette filler. (See Appendix D, pp. 127-28.)

Blood Sampling

In the elementary and secondary schools, written parental permission should be obtained for taking blood for testing from an individual student. Antiseptic techniques should be used. Use 70% alcohol on the skin where the blood is to be obtained both before and after the sample is taken. Finger tips are good sources. Use individually packaged sterilized lancets. Never reuse a lancet. Discard immediately after use. Never take more than a drop or two from an individual student. If larger amounts are needed for blood studies, outdated supplies from a blood bank may sometimes be obtained. Also, medical technician laboratories in hospitals and medical clinics can often provide larger quantities or "draw" some from volunteers from within or outside the class. It is best to break, after pressure and/or chemical sterilization, glass slides and containers that have been used in blood studies. Students should be cautioned that blood typing done in school laboratory settings should be checked with typing done by a certified technician. This is particularly desirable if the typing has been done by other students and if less frequent types are identified, such as AB and Rh negative.

Use of Glassware

The same precautions should be followed in the use of glassware in the biology laboratory as are used in the chemistry laboratory. Follow standard techniques in cutting and polishing glass tubing. The insertion of ordinary glass tubing or special forms such as thistle tubes through rubber stoppers can be dangerous unless the tube is lubricated slightly and some type of hand protection, such as a cloth or gloves, is used. Another technique that can be used for inserting glass tubing through stoppers is to place the next size larger cork-borer into the stopper hold. Push the tubing through the cork-borer and then withdraw the borer, leaving the tubing inserted in the stopper. Heat-resistant glass containers should be used when substances are being heated in them and particularly when materials with highly contrasting temperatures are added during the heating process. Glassware with any blemish (crack, nick, broken rim) should be discarded. It is tempting, at times, to use a clean beaker or flask on the storeroom shelf as a drinking cup. Avoid this temptation. Never use laboratory glassware for preparation or serving of food for human consumption.

Extractions

A number of substances are extracted from organisms and tissues through the use of solvents. A rather common activity is the extraction of chlorophyll from plant leaves using alcohol. Heating is necessary in order to do this effectively. Open flames should be avoided since the vapors, of course, are combustible. Hot plates should be used and a water bath will reduce the hazard. Most solvents are flammable and the use of water baths and hot plates

Safety in Biology Settings

with them are appropriate precautionary practices. Special attention needs to be given to the storage of these flammable solvents. Refrigerators are poor storage compartments because of danger of sparking from the electrical components present. When such types of chemicals require refrigerated storage, specially constructed refrigerators are needed. Many home-type refrigerators can be made explosion-safe by removing sources of ignition from the storage area. Relocating the thermostatic control to the outside and removing the light switch and any other switches from the storage area will usually accomplish this.

Wherever volatile chemicals are stored, adequate exhausting of vapors is necessary.

Heating Solids

Most solids should be heated in or over a water bath, depending upon the effect of water upon the chemical involved. Melting is usually done in a watch glass over the water bath. It is unwise to melt a solid in a test tube since it often melts and vaporizes within a rather narrow temperature range and will "spurt out" the end of the test tube. Never point the open end of a test tube toward anyone's face when any type of substance is being heated in it.

Sources of Heat

Reference has been made to sources of heat before. Avoid open flames, except when highly desirable, such as for sterilizing wire loops used in transferring samples of microorganism cultures. Alcohol lamps are inefficient sources of heat and the fuel requires special handling in storage and transfer. Candles produce deposits of carbon that are troublesome and the intensity of heat produced is often inadequate for the "experiment." If open flames are used, Bunsen burners or an adaptation of them for use on small portable bottles of propane or butane, should be used. Hot plates with solid heating surfaces are the most desirable overall source of heat for use in the biology laboratory.

USE OF MICROTOMES

Most secondary laboratories do not have a precision sliding or rotary microtome. When they do, careful instruction must be given for its use and only those students with approvals following instruction should be allowed to use it. (It is a good idea for such operation approval to be required for a number of pieces of equipment.) The common instrument used in elementary and secondary schools in preparing biological materials for light transmission microscopic examination is a razor blade. A single-edge blade of the non-injection type should be used. Some use a double-edge one with one edge heavily taped. This should be avoided. Do not overuse a single blade. As with scalpels, a dull blade is more dangerous than a sharp one properly used, and the materials prepared are much more satisfactory. If they do not interfere, lightweight gloves may be used when making thin sections with the razor blade to reduce injuries due to slippage of the blade.

DANGEROUS CHEMICALS

The same precautions in the usage of chemicals apply here as in the chemistry section. Reference should be made to that section. Careful labeling cannot be emphasized too much. Storage of particularly caustic substances, such as sulfuric and nitric acid, particularly if in large containers, should be on the floor of the storage space, or not more than one or two feet above the floor. The use of carbon tetrachloride should be avoided because of its effect upon the human liver. Picric acid, sometimes used as a biological stain and sometimes as an antiseptic, should never be used or present in secondary school laboratories because of its explosive nature. Some other reagents sometimes used in biology activities that require special attention in handling are Millon's reagent, phenol, colchicine, and Fehling's solution.

RADIOACTIVE MATERIALS

For dealing with radioactive hazards, refer to the chemistry section. Special regulations usually govern the use of radioisotopes; special attention should be given to these regulations.

SUMMARY

Most accidents in the biology classroom and laboratory and in the field can be avoided by regularly giving careful instruction about safety practices to follow and by regular and careful monitoring of the use or neglect of rational safety procedures. A person competent in first aid, first aid materials, and instructions should be always available. Accidents should be reported in detail and if the system does not have a standard report form, one should be developed. Again, the principle should always be emphasized that the total responsibility for safety in the classroom and laboratory must be shared by the teacher and student. To get the student to accept the appropriate share of that responsibility is one of the major objectives of instruction in the sciences.

REFERENCES

Hone, Elizabeth, et al., *A Sourcebook for Elementary Science*, 2nd ed. New York: Harcourt, Brace and World, 1971.
Kingsbury, John M., *Deadly Harvest: A Guide to Common Poisonous Plants*. New York: Holt, Rinehart and Winston, 1965.
Miller, David F., and Blaydes, Glenn W., *Methods and Materials for Teaching the Biological Sciences*, 2nd ed. New York: McGraw-Hill, 1962.
Morholt, Evelyn; Brandwein, Paul; and Joseph, Alexander, *A Sourcebook for the Biological Sciences*, 2nd ed. New York: Harcourt, Brace and World, 1966.
Safety First: Guidelines for Safety in the Science or Science Oriented Classroom in Delaware for Middle, Junior High, and Senior High Schools. Dover, Del.: Delaware Department of Public Instruction, 1976.
Safety in the Secondary Science Classroom. Washington, D.C.: National Science Teachers Association, 1978.

5

Safety in Chemistry Settings

CHEMICAL STORAGE

Proper storage of chemicals is an important aspect of any science safety program, providing (1) protection to the reagents from fire and environmental insults, (2) security against unauthorized access, (3) protection to the outside environment by restricting emission from stored chemicals.[1]

STORAGE CLASSIFICATIONS

Safe chemical storage requires an understanding of the hazards of the following classes of materials.[2]

1. Flammable liquids and solvents
2. Corrosive, hazardous, and reactive chemicals
3. Explosives
4. Compressed gases
5. Carcinogenic chemicals
6. Radioactive chemicals

Clearly there is some overlap in the above groups. Ideally, each class should be isolated from the others and there should even be some isolation within a class. This is not feasible in the average school science storage area, but safety steps can be taken.

Many science laboratories store reagents on shelves in alphabetical order to facilitate retrieval. Although it is a convenient system it almost surely results in incompatible chemicals being stored close to one another. For all groups, remember not to store liquids above eye level, and to store large containers near the floor.

One recommended storage pattern is printed below (Tables 5.1 and 5.2) with permission of Flinn Scientific, Inc. First the inorganic chemicals are separated from the organic chemicals. Then within the two major groups the chemicals are stored by families that are compatible with each other. Each numbered line represents a shelf with the chemical families that can be stored

---

1. *Safety in the School Science Laboratory*, U.S. Department of Health Education and Welfare, National Institute for Occupational Safety and Health, Cincinnati, Ohio 45226, 1977, p. 6-9.
2. *Hazardous Materials Safety Seminar*, National Hazards Control Institute, Starson Corporation, Stanton, N.J. 08885, p. 41-1.

Table 5.1. STORAGE PATTERN FOR INORGANIC CHEMICALS

Shelf Order (Top)

1. Sulfur, Phosphorus, Arsenic, P_2O_5
2. Halides, Sulfates, Sulfites, Thiosulfates, Phosphates, Halogens
3. Amides, Nitrates (except NH_4NO_3), Nitrites, Azides, HNO_3
4. Metals, Hydrides (store away from water)
5. Hydroxides, Oxides, Silicates, Carbonates, Carbon
6. Arsenates, Cyanides, Cyanales, HCN (store above acids)
7. Sulfides, Selenides, Phosphides, Carbides, Nitrides (store at eye level, away from water)
8. Borates, Chromates, Manganates, Permanganates
9. Chlorates, Perchlorates, Perchloric acid, Chlorites, Hypochlorites, Peroxides, H_2O_2 (store away from heat, near floor)
10. Acids (except HNO_3, HCN)

together in relative safety. These lists are not intended to be comprehensive, nor are they the only recommended methods of storage. But they can be conveniently used in the average high school lab.

Flammable Liquids and Solvents

Flammable and combustible liquids present a dangerous situation in science storage areas and labs. According to the National Fire Protection Association (NFPA), a flammable liquid fire occurs about every eight minutes. About 25 percent of all school and college fires start with ignition of a flammable liquid.

Table 5.2. STORAGE PATTERN FOR ORGANIC CHEMICALS

Shelf Order (Top)

1. Alcohols, Glycols, Amines, Amides, Imines, Imides
2. Hydrocarbons, Esters, Aldehydes
3. Ethers, Ketones, Ketenes, Halogenated Hydrocarbons, Ethylene Oxide
4. Epoxy compounds, Isocyanates
5. Sulfides, Polysulfides, Sulfoxides, Nitriles
6. Phenol, Cresols
7. Peroxides, Hydroperoxides, Azides
8. Acids, Anhydrides, Peracids (store away from other chemicals)

(Volatile ethers, hydrocarbons, etc. should be stored in an explosion-proof or explosion-safe refrigerator.)

General guidelines:
a. Storage areas should be cool and dry (between 55° and 80°F).
b. Avoid having ignition sources nearby (including no smoking).
c. Isolate flammable liquids from oxidizing agents.

A few terms will be defined here to provide a basis for understanding Occupational Safety and Health Administration (OSHA) and NFPA guidelines.

Flammable liquids are often characterized by their *flash point*. It is defined as the lowest temperature at which a liquid will give off flammable vapors in the vicinity of its surface sufficient to form a flammable mixture with air. The mixture at its flash point will ignite when exposed to a source of ignition. The flash point is empirical and varies with method of determination. Traces of other flammable liquids as contaminants can lower the flash point significantly. The flash point is a function of the vapor pressure of the liquid and of the lower flammable limit.

The *flammable limit* or explosive range is expressed in percentages by the volume of fuel vapor in air. It is the range in concentration over which a particular vapor or gas mixture with air will burn when ignited. The range is indicated by the lower flammable limit (LFL) and the upper flammable limit (UFL), which are determined at normal atmospheric temperature and pressure. Some references use the terms LEL (lower explosive limit) and UEL (upper explosive limit).

The ignition temperature is the minimum temperature to which flammable liquid vapor must be heated to initiate self-sustained combustion independent of the original heat source. Another parameter often considered with flammable liquids is the *auto-ignition temperature*. It is the temperature at which the liquid will self-ignite and sustain combustion in the presence of a spark or flame. These values are influenced by size, shape of the heated surface, heating rate, and oxygen concentration. One might expect that the ignition temperature will not be commonly obtained in a storage area. However, it takes only a static spark lasting a few thousandths of a second or a light bulb (surface temperature can reach several hundred degrees) to initiate a fire or explosion.

Generally, it is the flash point that is used as an indication of the fire hazard of the material. The lower flammable limit and evaporation rate are also important, but a standard definition of a combustible liquid is that it is any liquid with a flash point between 100° and 200°F. (Note that there are compounds with a flashpoint below 100°F, too.) Table 5.3 lists the properties of a few flammable liquids commonly found in secondary science labs. Table 5.4 gives the relationship of the flash point and boiling point to the NFPA fire hazard rating.

Table 5.5 gives examples from each class of common flammable liquids. Table 5.6 lists the maximum allowable size of containers for storage. Smaller containers in the instructional lab itself are recommended. Obviously the glass container offers the least protection and the safety can the most. Since the safety can is designed to release solvent vapor pressure, it will not rupture even in a fire, and is the storage container to choose if at all possible. Metal cans are not as good as safety cans but are still much better than glass bottles frequently used in school labs. In a fire a metal can will split open at a seam, discharging solvent into the fire. If flammable liquids must be stored in glass containers, protect the bottle from breakage, and use a suitable bottle carrier for transport.

The NFPA publishes standards which govern fire protection in educational institutions. The NFPA code (No. 30, Section 44) states that the storage of flammable liquids shall be limited to that required for maintenance, demonstration, treatment, and laboratory work. The code establishes the following storage provisions for flammable liquids:

Table 5.3. PROPERTIES OF SOME FLAMMABLE LIQUIDS FOUND IN SCIENCE LABS[3]

Material	NFPA Flammability Hazard	Flash Point(F)	Flammable Range (in Air)	Auto Ignition Temperature(F)	Vapor Density (Air=1)	Evaporation Rate (Ether=1000)	Boiling Point(F)	Vapor Pressure @ 20°C
Acetic acid	2	109°	5.4-16%	869°	2.1	--	245°	--
Acetone	3	0°	2.1-13%	869°	2.0	476	134°	184.5
Aniline	2	158°	1.3-?%	1139°	3.2	3	364°	0.42
Carbon disulfide	3	-22°	1.3-50%	194°	2.6	556	115°	298.0
Chlorobenzene	3	84°	1.3-7.1%	1184°	3.9	80	270°	8.8
Ether	4	-49°	1.9-36%	320°	2.6	1000	95°	442.0
Gasoline (100 Octane)	3	-36°	1.4-7.4%	--	3-4	--	100-400°	--
Hexane	3	-7°	1.1-7.5%	437°	3.0	435	156°	120.0
p-Xylene	3	--	1.1-7.0%	986°	3.7	71	281°	7.1

3. *Safety in the School Science Laboratory*, p. 6-3.

Table 5.4. DOT AND OSHA DEFINITIONS FOR FLAMMABLE LIQUIDS

Class	Criteria	NFPA Fire Hazard Rating
IA	Boiling point below 100°F (Flash point below 73°F)	4
IB	Boiling point at or above 100°F (Flash point below 73°F)	3
IC	Boiling point not considered (Flash point between 73°F and 100°F)	3
II	Flash point 100-140°F	2
IIIA	Flash point 140-200°F	2
IIIB	Flash point above 200°F	1

a. No container shall exceed a capacity of 1 gallon.
b. Not more than 10 gallons of flammable or combustible liquid shall be stored outside of a storage cabinet or storage room except in safety cans.
c. Not more than 25 gallons of flammable or combustible liquids shall be stored in safety cans outside a storage room or storage cabinet.
d. Quantities of flammable and combustible liquids in excess of those set forth in this section shall be stored in an outside storage room or storage cabinet.
e. Not more than 60 gallons of flammable or 120 gallons of combustible liquids may be stored in a storage cabinet.

Commercially available OSHA approved storage cabinets are equipped with locks and can be exhausted to prevent accumulation of explosive or toxic vapors. Construction specifications for storage rooms and use of commercially available flammable storage cabinets are discussed in the NFPA Fire Codes and should be consulted if lab modifications are being considered. (See Appendix D, p. 124.)

Corrosive, Hazardous, and Reactive Chemicals

A *corrosive chemical* is any liquid or solid that causes steel (SAE 1020 alloy) at 130°F to corrode at a rate of greater than 0.25 inches per year or causes destruction of human skin tissue.

General storage guidelines for corrosive chemicals[4]:
a. Store in a cool, dry, well-ventilated area away from direct sunlight.
b. Storage area should be constructed of noncorroding materials or covered with acid fume resistant paint.
c. Isolate them from reactive chemicals.
d. When possible, store corrosive liquids in their original shipping container, with label showing.
e. Store with label turned so others can see it without touching the container.
f. Keep the storage area locked if necessary.
g. Have appropriate rubber apron, gloves, and face shield nearby.

---

4. Ibid., pp. 41-7, 41-8.

Table 5.5. EXAMPLES OF COMMON FLAMMABLE LIQUIDS

	Flash Point(F)	Boiling Point(F)
Class I-A--Very Flammable		
acetaldehyde	−36	70
ether	−49	95
n-Pentane	−56	97
methylamine	33	21
Class I-B--Flammable		
acetone	0	133
benzene	12	176
t-Butyl alcohol	50	180
carbon disulfide	−22	115
cyclohexane	0	177
cyclohexene	−20	181
ethyl alcohol	54	174
n-heptane	25	209
n-hexane	−8	156
methyl alcohol	54	149
petroleum ether	50	
toluene	39	231
Class I-C--Flammable		
n-Butyl alcohol	84	
n-Propyl alcohol	77	
trichloroethylene	90	
xylene	81	
phenol red solution	77	
Class II--Flammable		
acetic acid	135	
n-pentanol	124	
Class III-A		
analine	158	
benzaldehyde	144	
butyric acid	171	
kerosene	153	
Class III-B		
glycerol	320	
olive oil	437	
Nonflammable		
carbon tetrachloride		
ethyl bromide		
No Flash Point		
chloroform		
methyl iodide		

Table 5.6. MAXIMUM ALLOWABLE SIZE FOR STORAGE

Class	Glass	Metal	Safety Can
IA	1 pt.	1 gal.	2 gal.
IB	1 pt.	5 gal.	5 gal.
IC	1 gal.	5 gal.	5 gal.
II, III	1 gal.	5 gal.	5 gal.

Safety in Chemistry Settings

    h. Special storage areas for especially hazardous materials should be established (for example, hydrofluoric acid and bromine).
    i. Spill clean-up agents should be located nearby.
    j. Store corrosive liquids below eye level.

A *hazardous chemical* is a substance that has been determined to be capable of posing unreasonable risks to health, safety, or property. The term hazardous material can have very broad meaning and the U.S. Department of Transportation has a general classification system used for shipping, shown below:

Class No.	Description
1	Explosives
2	Compressed gases
3	Flammable liquids
4	Flammable solids, water reactive substances
5	Oxidizing materials
6	Poisons
7	Radioactive materials
8	Corrosive materials
9	Materials not covered above

Here we are specifically interested in materials that commonly cause a human threat. For example, sodium hydroxide is highly hazardous and widely available in science labs, but generally not given proper respect. It can cause permanent irreversible eye injury in as little as 30 seconds after contact. Destruction of the cornea starts within 10 seconds after contact. Other common hazardous chemicals in the science lab are potassium chlorate and calcium carbide.

The Matheson, Coleman, Bell *MCB Chemical Reference Manual*[5] lists the following as hazardous or potentially hazardous chemicals. This list is not all-inclusive.

CLASSIFICATION OF 22 HAZARDOUS OR POTENTIALLY HAZARDOUS CHEMICALS

Acid Chlorides
Alkali Metals and Alkoxides
Aromatic Amines
Aromatic Halogenated Amines and Nitro Compounds
Aromatic Nitro Compounds
Azides
Carbon Disulfide
Caustic Alkalies
Chlorohydrins
Chlorosulfonic Acid
Heavy Metals

Chromium Trioxide
Cyanides and Nitriles
Ethers
Halogenated Hydrocarbons
Hydrazine and its Derivatives
Inorganic Amides
Mercury and its Derivatives
Oxalic Acid and Oxalates
Perchlorates
Peroxides
Phosphorus

Guidelines for storing hazardous chemicals:
    a. Know which chemicals in your storage area are hazardous. Label properly!

---

5. *MCB Chemical Reference Manual*, Vol. 2, Matheson, Coleman and Bell, 2909 Highland Avenue, Norwood, Ohio 45212, 1976, p. 308.

b. Know the specific health hazards involved.
   c. Know how to minimize exposure to you and your students (gloves, apron, and face shield should be nearby).
   d. Instruct all students about the potential hazards to instill proper respect.
   e. Minimize or eliminate storage of hazardous materials.
   f. Isolate hazardous materials from flammables and reactives.
   g. See guidelines above for applicable corrosive chemical rules.

Reactive chemicals will be limited here to two groups:
   a. Water and Air Reactive Chemicals. This includes alkali metals (sodium, potassium), strong acids, strong bases, acid anhydrides, carbides, and hydrides. They should be stored in a waterproof area away from other chemicals.
   b. Acid Sensitive Chemicals. This includes alkali metals, alkaline hydroxides, carbides, and cyanides. They will, in contact with acids or acid fumes, liberate flammable or toxic gases and heat; store them in a protected area.

Explosives

Unstable compounds that may decompose explosively are not uncommon in school science labs. Be cautious in storing peroxides (especially benzoyl peroxide), chlorates, perchlorates, azides, diazos, metal picrates, and picric acid.

Picric acid is sold in bottles of moist crystals, stabilized with 10 percent to 20 percent water. After they are opened, they should be inspected every six months for the presence of water. Water should be added if necessary and the contents of the bottle should be disposed of after two years.[6] The Iowa Department of Public Instruction recently sent picric acid disposal information to all school superintendents in the state; copies are available.

Ammoniacal silver nitrate should be prepared for use when needed and *should not be stored*.

Benzoyl peroxide bottles have exploded when being opened, while others have exploded upon standing. Apparently, if the chemical is stored in a cool, dark place there is little risk of explosion.[7]

The following functional groups are prone to instability (note the frequent occurrence of nitrogen):

$-O-O-$	(peroxide)	$-N=O$	(nitroso)
$-NO_2$	(nitro)	$-ONO_2$	(nitrate ester)
$-N=N-$	(Azo)	$-NHNO_2$	(nitramine)
$-N=$	(imine, etc.)	$=N-NO_2$	(nitroamine)
$-N_3$	(azide)		

Compressed Gases

A compressed gas is defined by the Department of Transportation (DOT) as "any material having in the container an absolute pressure exceeding 40 psi at $70°F$...." They tend to be more of a storage (and handling) problem than

---

6. "Picric Acid Provokes Alarm in Schools," *Chemical and Engineering News* 57 (February 26, 1979): 7-8.
7. *Chem 13 News*, No. 104 (April 1979): 8.

liquids and solids because of pressure, diffusivity, and low flash points--for flammable gases.

General guidelines:
a. Inspect gas cylinders upon receipt to insure that they are in satisfactory condition (look for rusting around the neck and valve).
b. Cylinders should always be stored with the protective screw-cap in place.
c. Keep the number of stored cylinders to a minimum.
d. Move cylinders using a suitable hand truck.
e. When returning empty cylinders leave some positive pressure in the cylinder and close the valve.
f. Do not store noncompatible gases such as hydrogen and oxygen together.
g. Secure all cylinders.
h. Do not store cylinders near boilers, steam or hot water pipes, or any source of ignition.

The DOT provides specifications related to compressed gas cylinders, their construction, and performance. If you intend to store compressed gas, please read a good reference, such as the *CGA--Compressed Gas Handbook* for detailed information.

Carcinogens

In 1974, OSHA promulgated standards for 14 carcinogens. In 1975, vinyl chloride was added and in 1976 coke oven emission was added. The original 14 compounds have many synonyms and common or trade names. It is recommended that these compounds be removed from school science labs and, if possible, acceptable substitutes be found for the experiments. Only under unusual circumstances involving a certified teacher who is knowledgeable in the handling of carcinogens should their use be allowed.

The original group included[8]

Methyl chloromethyl ether (CMME). Other names include: chlorodimethyl ether and chloromethyl ether. Route of entry: inhalation and skin absorption. Exposure limit: 0.1 percent by weight or volume.

3,3'-Dichlorobenzidine (and its salts). Other names include: 4,4'-Dichlorobiphenyl and 4,4' Diamino-o, o'Dichlorobenzidine. Route of entry: skin absorption. Exposure limit: 1 percent by weight or volume.

Bis (chloromethyl) ether (BCME). Other names include: Chloro (chloromethoxy) methane, sym-Dichloromethyl ether, and bis-CME. Route of entry: inhalation. Exposure limit: mixtures containing 0.1 percent by weight or volume.

Beta-Naphthylamine (2-NA). Other names include: 2-aminonaphthalene and 2-naphthalamine. Route of entry: inhalation, ingestion, and skin absorption. Exposure limit: 0.1 percent by weight or volume.

Benzidine. Other names include: Fast Corinth Base B, p-Diaminodiphenyl, 2-Aminodiphenyl, p,p'-Bianiline, Benzidine dihydrochloride, Benzidine sulfate, and 4,4'-Diaminobiphenyl. Route of entry: skin absorption. Exposure limit: mixtures containing 0.1 percent by weight or volume.

4-Aminodiphenyl. Other names include: 4-ADP, PAB, Biphenyline, p-Phenylaniline, Xenylamine, p-Aminobiphenyl, and p-biphenylamine. Route of entry:

---

8. *Carcinogens--Regulation and Control*, U.S. Department of HEW, NIOSH, Publication No. (NIOSH) 77-205, Cincinnati, Ohio 45226, 1977, pp. 32-48.

ingestion, inhalation, and skin absorption. Exposure limit: mixtures containing 0.1 percent by weight or volume.

Ethyleneimine (IEI). Other names include: Azirane, Azacyclopropane, Aziridine, Dimethylenimine, and Dihydroazirine. Route of entry: inhalation and skin absorption. Exposure limit: mixtures containing 1 percent by weight or volume.

Beta-Propiolactone (BPL). Other names include: 2-oxetanone, propiolactone, B-lactone hydrocrylic acid, 3-Hydroxypropionic acid lactone, and propanolide. Route of entry: inhalation and skin absorption. Exposure limit: mixtures containing 1 percent by weight or volume.

2-Acetylaminofluorene. Other names include: AAF, FAA, 2-Fluorenylacetamide, 2-Acetamidofluorene, and N-Acetylaminophenathrene. Route of entry: inhalation and skin absorption. Exposure limit: mixtures containing 1.0 percent by weight or volume.

4-Demethylaminoazobenzene (DAB). Other names include: dimethyl yellow, methyl yellow, benzeneazodimethyl analine, and N,N-dimethyl-p-phenylazo analine. Route of entry: ingestion, inhalation, and skin absorption. Explosure limit: mixtures containing 1 percent by weight or volume.

N-Nitrosodimethylamine (DMN). Other names include: Nitrous dimethylamide, N,N-Dimethylnitrosoamine, and Dimethylnitramine. Route of entry: inhalation and skin absorption. Exposure limit: mixtures containing 1 percent by weight or volume.

4-Nitrobiphenyl (4-NBP). Other names include: 4-Nitrodiphenyl, p-Nitrobiphenyl, p-Nitrodiphenyl, PNB. Route of entry: inhalation and skin absorption. Exposure limit: 1 percent by weight or volume.

Alpha-Naphthylamine (1-NA). Other names include: Naphthylamine, Fast Garnet Base B, 1-Aminopaphthalene, Naphthalidam, and Naphthalidine. Route of entry: inhalation, skin absorption, and ingestion. Exposure limit: mixtures containing 1.0 percent by weight or volume.

4,4'-Methylene bis (2-chloroaniline). Other names include: MOCA and DACPM. Route of entry: inhalation and skin absorption. Exposure limit: mixtures containing 1.0 percent by weight or volume.

Vinyl Chloride (VC). Other names include: chloroethylene, chlorethene, VCM, and monochloroethene. Route of entry: inhalation. Exposure limit: 1 ppm as 8-hour time weighted average.

In addition to these, a number of other compounds have been under close scrutiny. The following compounds are commonly found in school labs and appear in a 1978 *Chemical and Engineering News*[9] list of tentative carcinogens. It is recommended that if you have no very strong need for them, you eliminate them from your inventory.

<u>Tentative Carcinogen List</u>

acetamide	carbon tetrachloride
acetic acid, chromium (3+) salt	chloroform
acetic acid, lead (2+) salt	chromic acid, lead (2+) salt
arsenic	chromium
arsenic trioxide	lead acetate (11), trihydrate
asbestos	lead chromate (VI) oxide
benzene	nickel
cadmium	nickel (II) acetate (1:2)
cadmium chloride	tannic acid

---

9. "OSHA Issues Tentative Carcinogen List," *Chemical and Engineering News* 56 (July 31, 1978): 20-22.

OSHA is currently developing regulations to deal with other potential carcinogens. New policy establishes two broad categories for *occupational carcinogens--confirmed* (Category I) and *suspected* (Category II). Benzene is now considered as a Category I carcinogen by OSHA.

Radioactive Materials

These are any material or combination of materials that spontaneously emit ionizing radiation having a specific activity greater than 0.002 microcuries per gram. A later section of this manual is devoted entirely to radioactivity, but please note here that no radioactive source should be stored near reactive chemicals.

STORAGE RECOMMENDATIONS
Hood Storage of Reagents[10]

Although there is a need for ventilated storage, fume hoods should not be used for storage of chemical reagents for a number of reasons:

a. Many fume hoods are not designed to run 24 hours a day.
b. Stored chemicals waste working space in the hood.
c. In case of a fire or explosion in the hood, the stored materials would only complicate the situation.
d. Labels often are lost from bottles stored in a hood due to the frequency of corrosive environments.
e. Hoods provide no security for the reagents.

Food in Refrigerators

Food for human consumption should never be allowed in refrigerators used to store chemical reagents. The chance of contaminating the food is too great.

Flammable Reagents in Refrigerators

Standard refrigerators are often found in science laboratories but never should be used to store flammable or volatile chemical reagents. Standard refrigerators have several potential ignition sources: exposed electrical connections, thermostat, defrost control, defrost timer, defrost heater, solenoid, light, light switch, pilaster heater, motor overload switch, compressor terminals, and condenser fan.[11]

There are two alternatives, *explosion-proof* refrigerators and *explosion-safe* refrigerators. Explosion-proof refrigerators must be used if explosive or flammable vapors are present (Class I--see the section on flammable liquids for definitions). The explosion-proof unit has all exterior controls enclosed in explosion-proof housing and wiring is sealed in an explosion-proof conduit. The compressor is hermetically sealed to protect it from flammable vapors. The initial purchase price and proper installation of the explosion-proof refrigerator is relatively costly. Furthermore, they are designed for use where the entire area is explosion-proof (light switches, hot plates, etc.)

An explosion-safe refrigerator would be a suitable alternative for most school labs. Most home refrigerators can be made explosion safe simply by removing all sources of ignition from the box (light and door switch). (See Appendix D, pp. 116-21.) on explosion-safe refrigerators. These refrigerators are

---

10. *Safety in the School Science Laboratory*, pp. 7-13, 7-14.
11. Arthur M. Stevens, "Flammable Liquids in the Laboratory," *American Laboratory* 10 (December, 1978): 75-77.

not to be confused with explosion-proof refrigerators. Explosion-proof refrigerators have no internal or external source of ignition. All motors, switches, etc., are nonsparking. If explosive mixtures of flammables are contained in the room atmosphere, explosion-proof refrigerators should be used. However, all other equipment in the room would also have to be explosion proof, including lights and switches.)

Other Storage Suggestions

Many common lab chemicals require special storage consideration because of low melting point, deliquescence, efflorescence, or the prevention of caking, mold growth, or chemical decomposition. The Mallinckrodt Chemical catalog[12] recommends keeping a number of chemicals cool and dry, in tightly closed containers, including: acetaldehyde, albumen egg scales, aluminum chloride, ammonium carbonate, ferric ammonium sulfate, monochloracetic acid, and potassium thiocyanate.

Mallinckrodt also lists a number of chemicals that should be protected from freezing. Examples of these are acetic acid (99.5 percent) which freezes at $60°F$ and benzene which freezes at $41°F$. Please consult the Mallinckrodt catalog for a more complete list.

LABORATORY MANAGEMENT

Purchasing Philosophy

Purchasing agents frequently buy the lowest price product and instructors often want to purchase large enough quantities of chemicals to avoid frequent ordering. These philosophies must change to comply with NFPA and OSHA regulations and, more importantly, to maintain a safe storage area at the school.

From a safety standpoint, chemical reagents should be purchased in the smallest quantities possible, consistent with the manner in which they are used. As a rule of thumb, it is suggested that a one semester supply of any hazardous chemical reagent would constitute a satisfactory supply.[13]

The shipping and storage container should also be considered when ordering. If a vendor offers both metal and glass containers, it is frequently worth the additional cost to pay for the metal container for flammable liquids.

It would also be a good practice to request the material safety data sheet from the supplier or manufacturer when ordering chemicals. It contains relevant information about the physical and toxicological properties of the chemical. It may also contain information about minor components or impurities that are lacking on the label.

Inventory Control Procedures

A school science laboratory should have an inventory system for all chemical reagents. A uniform system of recording purchase dates, receiving dates, quantities received, and quantities used is suggested. This will become increasingly important as the published lists of carcinogens and toxic chemicals grow.

All reagents upon receipt should be marked with the receiving date. This is most important for time sensitive chemicals, such as o-tolidine, diethyl ether, and anhydrous perchloric acid. The effect of age on o-tolidine is

---

12. Mallinckrodt Laboratory Chemicals and Plasticware Catalog, St. Louis, Missouri 63160, p. 282.
13. *Safety in the School Science Laboratory*, p. 6-11.

Safety in Chemistry Settings   53

innocuous; it simply won't function as a colorimetric reagent after it has been air oxidized. But ethyl ether can form an explosive peroxide upon air oxidation. Anhydrous perchloric acid is also capable of explosive decomposition and should not be kept if it discolors ($HClO_4$ should not even be on the inventory of most school labs).

Since detailed record keeping could become very time consuming to the instructor, consideration should be given to involving the students. This could be especially meaningful if the data sheet for each chemical listed the additional information of precautions in handling and toxicity data.

Safety Inspections

Regular safety inspections should be conducted for storage (and laboratory) areas to:

a. Locate potential problems.
b. Report findings.
c. Assure follow-up to eliminate the hazards. It is preferable to have a safety committee consisting of chemistry instructor, other science instructors, and a supervisor.

Sample safety checklists have been included on pp. 108-11. They may be used in developing more thorough lists for the specific needs of a particular laboratory or school. Copies of any safety report should go to the school administration, including specific corrective actions that should be taken (in writing). Nothing can be accomplished unless corrective action is taken.

REFERENCES

"Compressed Gases Can Be Dangerous--An Explosion Case History." (Slide/cassette tape audiovisual materials are on loan for duplication from: Joseph R. Songer, U.S. Department of Agriculture, National Animal Disease Center, P.O. Box 70, Ames, Iowa 50010.)

Fire Officer's Guide to Dangerous Chemicals. National Fire Protection Association, Boston, Mass.

Fire Protection for Laboratories Using Chemicals. National Fire Protection Association, Boston, Mass.

Fire Protection Guide on Hazardous Materials. National Fire Protection Association, Boston, Mass.

Fire Protection Handbook, 14th ed. National Fire Protection Association, Boston, Mass.

A Management Guide to Carcinogens, NIOSH Publication No. 77-205, Cincinnati, Ohio, 1977.

Manufacturing Chemists Association, *Guide to Safety in the Chemical Laboratory*. New York: Van Nostrand Reinhold, 1972.

Meidl, James H., *Explosive and Toxic Hazardous Materials*. Beverly Hills, Calif.: Glencoe Press, 1970.

_____, *Flammable Hazardous Materials*, Fire Science Series. Beverly Hills, Calif.: Glencoe Press, 1970.

Meir, G. D., *Hazards in the Chemical Laboratory*. London: Chemical Society, 1977.

Meyer, E., *Chemistry of Hazardous Materials*. Englewood Cliffs, N.J.: Prentice-Hall, 1977.

Steere, N. V., *CRC Handbook of Laboratory Safety*, 2nd ed. Cleveland, Ohio: Chemical Rubber Co., 1971.

The following National Fire Protection Association standards contain valuable information: Fire Prevention Code NFPA 1; Flammable and Combustible Liquids Code NFPA 30; Indoor General Storage NFPA 231; Outdoor General Storage NFPA 231A; Electrical Equipment in Hazardous Locations, Purged Enclosures, NFPA 496.

HANDLING OF CHEMICALS

Mishandling of chemicals can result in burns, poisoning, fires, and explosions. This portion of the manual deals with several dimensions of chemical handling, including: (a) general suggestions for appropriate lab behavior; these are "common sense" suggestions that *students* should learn early in the year; (b) more specific suggestions that apply to commonly performed school lab experiments; (c) procedures for handling corrosive solids, liquids, and gases; (d) incompatibility of specific compounds or groups of compounds; (e) chemical research, and (f) chemical dispensing.

BEHAVIOR IN THE LABORATORY[1]

1. Do not eat food in the lab. There is always a chance of accidental ingestion of chemicals. Never eat or drink from lab glassware. Remember that the majority of chemicals found in the lab can be classified as poisonous. Wash your hands after any experiment has been performed.
2. Do not smoke in the lab. It is both a source of ignition for flammables and a route for accidental chemical ingestion.
3. Wear appropriate eye protection during all chemical experimentation and wear other protective equipment as directed. Wear appropriate personal protective clothing to avoid possible skin absorption of chemicals.
4. Do only experiments that are authorized by the teacher.
5. If acids, bases, or corrosive materials are spilled on skin or eyes, immediately flush with water for 15 minutes. Seek medical attention. Do not attempt to neutralize acid or base spills on the skin or eyes.
6. If acids, bases, or other hazardous materials are spilled on clothing, notify the instructor immediately.
7. If acids, bases, or other hazardous materials are spilled on counters, floor, cabinets, etc., notify the instructor immediately.
8. Do not touch chemicals unless directed to do so.
9. Do not taste chemicals, even if a lab manual suggests it.
10. Do not use the sink to discard matches, filter paper, or insoluble solids.
11. Have waste jars available for solids that should not be placed in the sink, being careful not to mix incompatible chemicals.
12. Check the label on reagent bottles twice before removing any of the contents.
13. Do not return unused chemicals to stock bottles.
14. Use a fume hood when dealing with volatile or toxic fumes.
15. Never point the open end of a test tube being heated at yourself or anyone else.
16. Never add water to concentrated sulfuric acid. Slowly add the acid to water with constant stirring.
17. Teachers and students must be shielded from demonstrations involving po-

---

1. *STOP--Safety First in Science Teaching*, Division of Science, North Carolina Department of Public Instruction, 1977, pp. 5-6.

tentially explosive substances.
18. Know the location of fire extinguishers, fire blanket, safety shower, eye wash, and fire alarm box.
19. Report to the teacher any personal injury sustained--burns, scratch, cut, or corrosive chemical spill--no matter how trivial it may appear. This should become a part of the school records.
20. It is recommended that spill clean-up materials be located in the lab to accommodate acid, base, and flammable solvent accidents.
21. Test for odor of chemicals by wafting the hand over the container and sniffing cautiously.
22. Never use mouth suction on a pipette filled with a chemical reagent. Use a suction bulb.
23. Always work in a well-ventilated area.
24. Handle hazardous and flammable liquids over a pan to contain spills.
25. Keep flammable liquids away from flames, hot plates, and high temperature devices.

PROCEDURES FOR COMMONLY PERFORMED EXPERIMENTS[2]
1. When collecting gas by water displacement, remove the delivery tube from the generating flask prior to removing the heat so that water will not draw back into the hot generator.
2. Wrap gas generators in a towel. Keep generator size to a minimum.
3. In glass bending operations, be cautious to avoid burns.
4. Store dry ice in a styrofoam chest. Do not allow dry ice to touch the skin. Do not place dry ice in airtight bottles.
5. Store and cut white phosphorus under water. Substitute red phosphorus for white when possible. (Eliminate white phosphorus from your inventory entirely if possible.) Do not dump elemental phosphorus in waste jars; burn residues in a hood.
6. After opening containers of metallic sodium and potassium, store them under kerosene. Do not allow contact with skin.
7. Do not do perchloric acid digestions in your fume hood unless it is specifically designed to accommodate this operation.
8. If you discover an old can of ether or other peroxide forming compound, treat it like a bomb. Seek expert advice.
9. Substitute methylene dichloride (CH_2Cl_2) for carbon tetrachloride in lab experiments whenever possible.
10. Avoid using ammonium dichromate in "volcano models." It is toxic, flammable, and potentially explosive in the presence of some organic compounds. Substitute another compound.
11. When generating toxic gases in the lab (chlorine, sulfur dioxide, nitrogen oxides), make only very small quantities. Use the hood.
12. When handling mercury:
    a. Use it only in well-ventilated areas.
    b. Store mercury in tightly closed containers.
    c. Avoid contact with skin of mercury and its compounds.
    d. Do not flush mercury or its compounds down the drain.
    e. Avoid decomposition of mercuric oxide as an experiment.
    f. Take extreme care to avoid spillage. Do the experiment in or over a tray.
    g. Remember that mercury has no odor.

---

2. *Science Classroom and Laboratory Procedures*, Waterloo Community Schools, Waterloo, Iowa, 1968, p. 24.

13. When handling ether:
    a. Never perform a distillation or evaporation without testing for the presence of peroxides. One test follows: Shake 10 ml of ether for one minute with 1 ml fresh 10 percent potassium iodide in a 25 ml glass-stoppered cylinder of colorless glass that is protected from the light. View transversely against a white background. The absence of peroxide is indicated by the lack of any color in either layer.
    b. Avoid the use of open flames around the ether.
    c. A good rule is to discard containers of ether three months after opening; discard closed containers within one year of receipt.

CORROSIVE SOLIDS, LIQUIDS, AND GASES[3]

If a corrosive compound is called for in an experiment, first determine whether a less hazardous compound could be substituted. If not, be sure appropriate protective equipment is available. As a rule of thumb, corrosive solids are considered a less immediate hazard than corrosive liquids, which are in turn less an immediate hazard than corrosive gases.

Corrosive liquids are the most commonly used in the lab. Please note the following guidelines:

1. Use bottle carriers when moving glass bottles of dangerous liquids.
2. Keep spill control materials close to areas where corrosives are handled.
3. Use the fume hood when transferring corrosive liquids.
4. Never allow anyone to work alone in the lab, especially with corrosives.
5. Pour corrosive chemicals over a pan or tray to confine spills.

When handling gases observe the following: (See also the section in this book on storage of gases.)

1. Always use the proper regulator.
2. Recognize that compressed gas cyliners can become unguided missiles if the valve is sheared off. Do not move cylinders without the screw-on cap being in place.
3. Recognize that some compressed gases will support combustion (oxygen, nitrous oxide, chlorine, fluorine).
4. Some compressed gases are flammable (hydrogen) and have very low flash points.
5. Some compressed gases are corrosive or toxic (nitrogen dioxide).

If your lab does intend to handle compressed gases, please consult a good reference for more information before starting.

INCOMPATIBLE CHEMICALS

There are so many potentially hazardous chemical combinations that it would be difficult for even a veteran chemistry teacher to know them all. Accidents will best be avoided by not allowing any unauthorized experimentation.

---

3. *Safety in the School Science Laboratory*, p. 6-3.

Table 5.7. INCOMPATIBLE CHEMICALS

Chemical	Should not come in contact with
Acetic acid	Nitric acid, peroxides, permanganates, ethylene glycol, hydroxyl compounds, perchloric acid, or chromic acid
Acetone	Concentrated sulfuric and nitric acid mixtures
Acetylene	Bromine, chlorine, fluorine, copper tubing; as well as silver, mercury, and their compounds
Alkali metals	Carbon tetrachloride, carbon dioxide, water, halogens
Alkaline metals (powdered aluminum or magnesium)	Carbon tetrachloride or other chlorinated hydrocarbons, halogens, carbon dioxide
Ammonia, anhydrous	Mercury, hydrogen fluoride, and calcium hypochlorite, chlorine, bromine
Ammonium nitrate	Acids, flammable liquids, metal powders, sulfur, chlorates, any finely divided organic or combustible substance
Aniline	Nitric acid and hydrogen peroxide
Bromine, chlorine	Ammonia, petroleum gases, hydrogen, sodium, benzene, finely divided metals
Carbon, activated	Calcium hypochlorite and all oxidizing agents
Chlorates	Ammonium salts, acids, metal powders, sulfur, and finely divided organic or combustible substance
Chromic acid	Glacial acetic acid, camphor, glycerin, naphthalene, turpentine, lower molecular weight alcohols, and many flammable liquids
Copper	Acetylene and hydrogen peroxide
Flammable liquids	Ammonium nitrate, chromic acid, hydrogen peroxide, sodium peroxide, nitric acid, and the halogens
Hydrocarbons (propane, benzene, gasoline)	Fluorine, chlorine, bromine, sodium peroxide, and chromic acid
Hydrofluoric acid	Ammonia (aqueous or anhydrous)
Hydrogen peroxide	Most metals and their salts, alcohols, organic substances, any flammable substance
Hydrogen sulfide	Oxidizing gases, fuming nitric acid
Iodine	Acetylene, ammonia, hydrogen
Mercury	Acetylene, ammonia

Table 5.7 *(continued)*

Chemical	Should not come in contact with
Nitric acid (concentrated)	Acetic acid, hydrogen sulfide, flammable liquids and gases, chromic acid, aniline
Oxygen	Oils, grease, hydrogen, flammable liquids, solids, and gases
Perchloric acid	Acetic anhydride, bismuth and its alloys, alcohols, paper, wood, and other organic materials
Phosphorus pentoxide	Water
Potassium chlorate	Sulfuric and other acids, any organic substance
Potassium permanganate	Sulfuric acid, glycerine, ethylene glycol
Silver	Acetylene, ammonia compounds, oxalic acid, and tartaric acid
Sodium peroxide	Ethyl or methyl alcohol, glacial acetic acid, carbon disulfide, glycerine, ethylene glycol, ethyl acetate
Sulfuric acid	Potassium chlorate, potassium perchlorate, potassium permanganate, similar compounds of other light metals

The generalizations in Table 5.7 should be applicable most of the time. Please consult one of the publications at the end of this section for more data on chemical reactions and incompatible chemicals.[4,5]

Figure 5.1 should serve as a reference for those who plan to mix chemicals with which they are not familiar.

CHEMICAL RESEARCH

Frequently accidents occur in the science laboratory simply because neither the instructor nor the student knows or is able to anticipate the effects of a particular chemical combination. This is a not uncommon situation even among highly experienced chemists.

A primary goal of any school science safety program should be to minimize the frequency and severity of accidents that result from a lack of knowledge. Data on chemical reactions and incompatible chemical compounds and elements have been collected in several publications. This information should be readily available to all high school science instructors and should be used by them whenever necessary, so that bodily injuries and property damage resulting from unexpected chemical reactions can be eliminated as much as possible.

NOTE: A copy of the *Manual of Hazardous Chemical Reactions* is recommended for high school science departments. The 470-page book (491M) is available for $5.00 (1977 price) from the National Fire Protection Association, 470 Atlantic Avenue, Boston, MA 02210.

When starting any research, the experimenter should:

---

4. *Science Classroom and Laboratory Procedures*, pp. 27-28.
5. *MCB Chemical Reference Manual*, Vol. 2, Matheson, Coleman and Bell, 2909 Highland Avenue, Norwood, Ohio 45212, 1976, pp. 329-330.

Safety in Chemistry Settings   59

COMPATIBILITY CHART[6]

1. Acid Anhydrides	1															
2. Alcohols	X	2														
3. Aldehydes	X		3													
4. Amines	X	X	X	4												
5. Aromatic Hydrocarbons					5											
6. Caustics	X	X				6										
7. Esters			X		X		7									
8. Ethers	X							8								
9. Halogenated Compounds			X	X					9							
10. Halogens		X	X		X	X	X			10						
11. Inorganic Acids		X	X	X	X	X	X	X	X		11					
12. Ketones		X	X		X					X	X	12				
13. Olefins										X	X	X	13			
14. Organic Acids			X	X	X						X			14		
15. Oxidizing Agents	X	X	X	X	X	X	X	X	X	X	X	X	X	X	15	
16. Saturated Hydrocarbons										X					X	16

X Represents potentially unsafe combinations

Read the chart first from left to right, and then down.

Fig. 5.1. Compatibility Chart. (*Hazardous Materials Safety Seminar*, pp. 41-11.)

1. Have knowledge of the violence of reactivity to be expected.
2. Have knowledge of the procedures to be used, and "good practices" to be followed.
3. Use adequate protective equipment.
4. Use small quantities of chemicals initially to determine the force of reactions.
5. Perform tests in a good fume hood.

CHEMICAL DISPENSING

   Dispensing of chemicals by authorized personnel has its advantages over the storing of chemicals on open shelves. It provides protection against unauthorized experiments, theft, misreading of similar labels, misplacing of reagents, security of hazardous chemicals, and contamination of stock reagents.

REFERENCES

DRAFT: "Guidelines for Laboratory Use of Chemical Substances Posing a Potential Occupational Carcinogenic Risk," by the Laboratory of Chemical Carcinogenic Safety Standards Subcommittee of the HEW Committee to Coordinate Toxicology and Related Programs.
For information on this draft write:
NCI (National Cancer Institute)
NIH (National Institutes of Health)
Bethesda, MD 20014
Contact: Emmett Barkley, Chair
Phone: Ms. Rhonda Rice 1-(919)-541-3506.

Handbook of Chemistry and Physics, 57th ed. Chemical Rubber Company, 1976.

Handbook of Hazardous Materials, 491M, National Fire Protection Association, Boston, Mass.
Hazardous Chemicals Data 1975, 49, National Fire Protection Association, Boston, Mass., 1975.
Kirk, R. E., and Othmer, D. F., *Encyclopedia of Chemical Technology*. New York: Wiley, 1970.
Lange, N. A., *Handbook of Chemistry*, rev. 11th ed. New York: McGraw-Hill, 1973.
Manufacturing Chemists Association, Inc., *Chemical Safety Data Sheets*. 1825 Connecticut Ave., N.W., Washington, D.C. (54 sheets available).
Manufacturing Chemists Association. *Guide for Safety in the Chemical Laboratory*, 2nd ed. New York: Von Nostrand Reinhold, 1972.
Matheson Gas Data Book. East Rutherford: Matheson Company, Inc., 1961.
Meidl, James H., *Flammable Hazardous Materials*. Beverly Hills, Calif.: Glencoe Press, 1970.
Merck Index, 9th ed. Rahway, N.J.: Merck and Company, 1976.
Meyer, E., *Chemistry of Hazardous Materials*. Englewood Cliffs, N.J.: Prentice-Hall, 1977.
Patty, F. A., *Industrial Hygiene and Toxicology*, 2nd ed. New York: Wiley, 1958.
Steere, N. V., *CRC Handbook of Laboratory Safety*, 2nd ed. Cleveland, Ohio: Chemical Rubber Co., 1971.

HEALTH HAZARDS

The majority of chemicals found in the chemical lab can be classified as poisonous. "There are no safe substances, only safe doses." The three common routes of entry of toxic substances into the body are by inhalation, ingestion, and absorption through the skin.

A number of special terms have been developed to deal with toxicity and industrial hygiene. Only a very limited discussion can be included here.

TERMS

TVL or Threshold Limit Value--the maximum value of permitted exposure over an eight hour work day with no ill effect. (Any compound with a TLV below 100 parts per million (ppm) is considered extremely hazardous.) The term PEL--Permissible Exposure Limit--is now used by OSHA for TLV's, and is often given in units of mg/m^3 (milligrams per cubic meter).

MAC--the maximum allowable concentration to which a person should be exposed.

TWA--the time weighted average representing an estimate of how much below the maximum exposure the concentration should be kept for safety to personnel working a 40 hour week in the area. (NIOSH TWA's are based on up to a 10 hour exposure unless otherwise noted.)

For information on irritants, asphyxiants, anesthetics, systemic poisons, particulate poisons, teratogenic effects, mutagenic effects, and synergistic effects, consult one of the toxicology or industrial hygiene books listed at the end of this section.

Safety in Chemistry Settings

TOXICITY

Since inhalation would be the most common way that toxic chemicals would enter the body in school science labs, a sample list[1,2] of upper limits of atmospheric concentrations for a few toxic chemicals is shown in Table 5.8.

Note that virtually every compound listed above is in the "extremely hazardous" classification (PEL below 100 ppm). Also keep in mind when working with compounds that are both hazardous and volatile that the vapor pressure as well as the PEL becomes significant in safety considerations.

It should be mentioned that TLV's (or PEL's) are subject to change. In 1957, the TLV for carbon tetrachloride was 25 ppm; now it is 10 ppm and NIOSH has recommended lowering it to 2 ppm for certain work environments.[3] Likewise, the PEL for chlorine is under scrutiny at present.

Benzene is a common lab chemical that has received a great deal of attention because of its link to leukemia in workers exposed over long periods. Acute toxicity symptoms include central nervous system manifestations, vertigo, convulsions, narcosis, depression, respiratory paralysis, and pulmonary edema. It can be absorbed through the skin as well as by inhalation.

Its PEL of 10 ppm is below the concentration at which the human nose can detect it. (Likewise, with carbon tetrachloride, the PEL is 25 ppm but it is not detected by smell until the concentration reaches 80 ppm.)

A sample calculation will graphically display the small quantities needed to reach dangerous concentrations. Benzene has a PEL of 10 ppm. Let's assume you are working in a good size lab, 30 ft x 40 ft x 10 ft. If one student were to allow 15 ml of benzene to evaporate into the lab (assuming it all stayed in the lab) the concentration of benzene in air would just exceed 10 ppm. (A good rule of thumb for benzene is, "If you can smell it, you're breathing too much of it.") Substitute toluene for benzene whenever possible.

Several other gases can be present in the air at dangerous concentrations *without detection*. These include carbon disulfide, mercury, carbon monoxide, dimethyl sulphate, hydrogen cyanide, methyl bromide, and ozone. Hydrogen sulfide so "paralyzes" the sense of smell that continued exposure goes undetected. Easily generated gases that are classified as corrosive are hydrogen chloride, ammonia, and formaldehyde.

Corrosive chemicals deserve special attention. They are hazardous in two ways: (1) by destructive action on living tissue such as direct chemical attack (dehydration or nitration), destruction of protein and body tissue, and disruption of the cell membrane, or (2) by reaction with materials producing toxic, flammable, or noxious products. On the skin, corrosives can cause skin burns and even charring. Inhalation can cause severe lung damage and if swallowed, corrosives can cause serious burns of the gastrointestinal tract from the mouth to the intestines.[4]

Phenol is an example of a frequently handled corrosive. The vapor can attack the lung tissue causing pulmonary edema. When it contacts the skin it can cause severe burns and penetrate the skin and act as a systemic stomach poison. It can cause death.

---

1. *Science Classroom and Laboratory Safety Procedures*, Waterloo Community Schools, Waterloo, Iowa, 1968, p. 31.
2. "Summary of NIOSH Recommendations for Occupational Health Standards," *Chem 13 News*, No. 110 (January 1980): 8-11.
3. *Hazardous Materials Safety Seminar*, National Hazards Control Institute, Starson Corporation, Stanton, N.J. 08885, 1978, p. 16-2.
4. "Skin: Function, Wounds, Treatment--Part 4," *Chem 13 News*, No. 106 (September 1979): 5.

Table 5.8. UPPER LIMITS OF ATMOSPHERIC CONCENTRATIONS

Compound	Current OSHA PEL	Health Effect Considered	Comments
Acetic Acid	10 ppm		
Acetic anhydride	5 ppm		
Ammonia	50 ppm; 8-HR TWA	Airway irritation	Eye damage
Aniline	5 ppm		
Benzene	10 ppm; 8-HR TWA 25 ppm acceptable ceiling 50 ppm maximum ceiling (10 min)	Blood changes	NIOSH recommends 1 ppm ceiling
Carbon Disulfide	20 ppm; 8-HR TWA 30 ppm acceptable ceiling 100 ppm maximum ceiling	Heart, nervous, and reproductive system effects	NIOSH recommends 10 ppm ceiling (15 min)
Carbon Monoxide	50 ppm; 8-HR TWA	Heart effects	
Carbon Tetrachloride	10 ppm; 8-HR TWA 25 ppm acceptable ceiling 200 ppm maximum ceiling (5 min in 4 hours)	Liver effects	
Chlorine	1 ppm; 8-HR TWA	eye/airway irritation	
Formaldehyde	3 ppm; 8-HR TWA 5 ppm acceptable ceiling 10 ppm maximum ceiling (30 min)	Irritation, lung effects	NIOSH recommends 1 ppm ceiling (30 min)
Hydrogen Sulfide	20 ppm acceptable ceiling 50 ppm maximum ceiling (10 min)	Irritation, nervous and respiratory systems	NIOSH recommends 10 ppm ceiling (10 min)
Iodine	1 ppm		
Mercury		Central nervous system and mental effects	
Methyl Alcohol	200 ppm TWA	Blindness, metabolic acidosis	
Nitric Acid	2 ppm; 8-HR TWA	Dental erosion, nasal/lung irritation	Hazardous liquid, eyes and skin
Nitrogen oxides NO, NO_2, N_2O_4	NO_2 5 ppm, 8-HR TWA	Airway effects	
Ozone	0.1 ppm		
Phenol	5 ppm; 8-HR TWA (skin)	Skin, eyes, central nervous system, liver and kidney effects	

Table 5.8 *(continued)*

Compound	Current OSHA PEL	Health Effect Considered	Comments
Sodium hydroxide	2 mg/m^3; 8-HR TWA	Airway irritation	Hazardous liquid, eyes and skin
Sulfur dioxide	5 ppm; 8-HR TWA	Respiratory effects	NIOSH recommends 0.5 ppm TWA
Sulfuric acid	1 mg/m^3; 8-HR TWA	Pulmonary irritation	Hazardous liquid, eyes and skin
Toluene	200 ppm; 8-HR TWA 300 ppm acceptable ceiling 500 ppm maximum ceiling (10 min)	Central nervous system depressant	

Other common lab chemicals that can penetrate the skin include methyl alcohol, butyl alcohol, benzene, carbon disulfide, aniline, and many pesticides. A list of prescribed skin absorption hazards has been published by OSHA (29 C.F.R. 1918.1000). Aniline and related compounds also cause dermatitis.

Mercury deserves a special niche in any chemical toxicology discussion. Its maximum safe concentration is 0.012 ppm. In a lab at room temperature with an open source of mercury, the mercury vapor concentration can rise to 1.84 ppm, 150 times the safe maximum. Since mercury is a cumulative poison in humans, extra care in keeping it in airtight containers and avoiding spillage is absolutely essential. (See also sections on handling and disposal for additional mercury information.)

CARCINOGENS

When OSHA first declared fourteen chemicals were carcinogenic, various State Departments of Labor around the country did surveys and found all fourteen chemicals were present in public schools. (Please refer to the earlier section on chemical storage for a list of carcinogens.)

These compounds are extremely toxic and may have effects at parts per billion or less. Moreover, the effects may not be manifested for years. It is recommended that these compounds be removed from the science lab. If they are absolutely essential for special projects which include strict personal supervision, then handling guidelines published by the government should be adhered to, including a regulated work area.[5,6]

Current OSHA policy defines a *potential occupational carcinogen* as "any substance or combination or mixture of substances which cause an increased incidence of benign and/or malignant neoplasms or a substantial decrease in the latency period between exposure and onset of neoplasms in humans or in one or more experimental mammalian species as the result of any oral, respiratory, or

---

5. *Carcinogens--Regulation and Control*, U.S. Department of HEW, NIOSH, Publication No. (NIOSH) 77-205, Cincinnati, Ohio 45226, 1977.

6. *Carcinogens--Working with Carcinogens*, U.S. Department of HEW, NIOSH, Publication No. (NIOSH) 77-206, Cincinnati, Ohio 45226, 1977.

dermal exposure, or other exposure which results in the induction of tumors at a site other than the site of administration. Exposure may be at any level.[7]

A 1979 NIOSH study[8] revealed that many types of gloves and other protective clothing widely recommended to protect against exposure to carcinogenic liquids appear to be ineffective when used for more than a few minutes. For example, neoprene rubber was the only widely recommended material that would prevent breakthrough of benzene for more than ten minutes.

---

7. "OSHA Rules Will Speed Carcinogen Regulation," *Chemical and Engineering News* 58 (January 28, 1980): 30.

8. "Protective Clothing Not So Good," *Chemical and Engineering News* 57 (April 2, 1979): 15.

REFERENCES

Casarett, L. J., and Doull, J., *Toxicology: The Basic Sciences of Poisons*. New York: Macmillan, 1975.

Christensen, H. E., and Luginbyhl, T., *Suspected Carcinogens: A Subfile of the NIOSH Toxic Substances List*. Rockville, Md.: U.S. Department of Health, Education and Welfare, Public Health Service, National Institute for Occupational Safety and Health, June 1975.

Christensen, H. E.; Luginbyhl, T. T.; and Carroll, B. S., *Registry of Toxic Effects of Chemical Substances*. Rockville, Md.: Department of Health, Education and Welfare, Public Health Service, National Institute for Occupational Safety and Health, June 1975.

Cusack, Michael, "Mercury--A Maddening Menace," *Science World* 19, No. 3 (Sept. 29, 1969): 10-12.

"Danger: School Science Projects," *Safety Newsletter*, September 1965, National Safety Council, 444 N. Michigan Avenue, Chicago, Ill. 60611.

Fawcett, H. H., "Health Aspects of Common Laboratory Chemicals," *Science Teacher* 17 (December 1966): 44-45.

Livingston, H. K., "Safety Considerations in Research Proposals," *Journal of Chemical Education* 41, no. 10 (1964): A785-89.

Manufacturing Chemists Association, *Guide for Safety in the Chemical Laboratory*, 2nd ed. New York: Van Nostrand Reinhold, 1972.

Manufacturing Chemists Association, Inc., *Chemical Safety Data Sheets*, 1825 Connecticut Ave., N.W., Washington, D.C. (54 sheets available).

Meidl, James H., *Explosive and Toxic Hazardous Materials*. Beverly Hills, Calif.: Glencoe Press, 1970.

Merck Index, 9th ed. Rahway, N.J.: Merck and Company, 1976.

Norwood, O. H., *Safety in Handling Hazardous Chemicals*. Manufacturing Chemists Association, 1971.

Olishifishi, Julian B., and McIlroy, Frank E., *Fundamentals of Industrial Hygiene*. Chicago, Ill.: National Safety Council, 1977.

Steere, N. V., *CRC Handbook of Laboratory Safety*, 2nd ed. Cleveland, Ohio: Chemical Rubber Company, 1971.

TLV Booklet, *Threshold Limit Values*, published yearly by the American Conference of Governmental Industrial Hygienists, P.O. Box 1937, Cincinnati, Ohio 45201.

"Warning about Carbon Tetrachloride," *Chemistry* 44, no. 6 (June 1971): 3.

Weast, R. C., *Handbook of Chemistry and Physics*, 56th ed. Cleveland: CRC Press, 1975-1976.

Safety in Chemistry Settings   65

CHEMICAL DISPOSAL

One of the most important aspects of the lab operation is disposing of waste chemicals. Toxic and hazardous wastes can be difficult to dispose of properly. It can be time consuming, costly, and have legal ramifications.

The U.S. Environmental Protection Agency (EPA) defines a hazardous waste as a waste which poses a threat to life and property.[1] Such wastes can poison, burn, and kill people and organisms. They may work their way into the food chain.

There are several reasons why disposal is necessary:

1. Chemical changes can occur during storage that make the chemical useless for the intended purpose.
2. The reagent may have lost its label and thus its identity.
3. A safer substitute may have been found to replace a hazardous or even carcinogenic compound.

DISPOSAL CHOICES

The EPA[2] lists the following disposal options for small batches of hazardous wastes, in order of preference:

1. Recycling or returning to the supplier.
2. Transporting to a hazardous waste management facility.
3. Using available lab equipment for treatment/disposal.
4. Disposing of materials in a municipal incinerator with permission of authorities.
5. Disposing of material in a landfill with permission of authorities.

Options 1 and 2 above are not particularly accessible to Iowa school science teachers, which leaves 3 as the best choice. Generally this would involve neutralizing the chemical or converting it to a less hazardous material or, in some cases, burning waste organics.

Small amounts of dilute acids, bases, or salt solutions may be flushed down the drain with large amounts of water. Volatile, corrosive, toxic, or insoluble salts should *not* be flushed down the drain.

There are so many waste disposal procedures that can be used "in house" that it is recommended that each chemistry lab teacher have a copy of the *Laboratory Waste Disposal Manual* published by the Manufacturing Chemists Association[3] or the *MCB Chemical Reference Manual*[4] or the *Aldrich Catalog Handbook of Fine Chemicals*.[5]

---

1. *Disposing of Small Batches of Hazardous Wastes*, U.S. Environmental Protection Agency, Report No. SW-562, 1976, p. 3.
2. Ibid., p. 8.
3. *Laboratory Waste Disposal Manual*, Manufacturing Chemists Association, 1825 Connecticut Avenue, N.W., Washington, D.C. 20009.
4. *MCB Chemical Reference Manual*, Vol. 2, Matheson, Coleman and Bell, 2909 Highland Avenue, Norwood, Ohio 45212, 1976. (The Safety Handbook starts on p. 307.)
5. *Aldrich Catalog Handbook of Fine Chemicals*, Aldrich Chemical Company, 940 W. St. Paul Avenue, Milwaukee, Wisconsin 53233, 1978. (The Waste Disposal Section is on pp. 1033-1034.)

As an example, assume you have some acetyl chloride you need to dispose of. You look up acetyl chloride in the alphabetical listing of the Aldrich Catalog and it states that disposal technique "I" should be used. You then look up "I" in the back of the catalog under Waste Disposal Procedures and read, "Carefully mix the acidic compound with dry sodium bicarbonate. Dilute slowly with water and wash down the drain with excess water."

The Manufacturing Chemists Association lists a number of compounds that can be dumped into landfills. Some that would commonly be found in school science labs are: calcium carbonate, calcium oxide, magnesium oxide, sulfur, and zinc oxide.

In some instances the chemical manufacturer or distributor of a given chemical needs to be contacted for information on waste disposal (or handling procedures and hazardous characteristics).

CHEMICAL SPILLS

Spill control is a universal problem in laboratories. In industry the most serious lab accidents, leading to loss of sight, burns, and the like, result from the accidental spilling of hazardous or corrosive chemicals.

You must be prepared to minimize the health and fire hazards associated with lab spills immediately. Allowing a quantity of carbon tetrachloride to simply evaporate as a clean-up technique could easily violate the OSHA permissible exposure limit and thus endanger your students. Likewise, soaking up a benzene spill with paper towels and placing them in a wastebasket can create a serious fire threat.

A recommended spill control method for acids, bases, and solvents involves the use of a chemically nonreactive absorbent material. By converting the spill to a solid form, it can more easily be picked up so that neutralization can be conducted under controlled conditions.

Materials that can be used include diatomaceous earth, vermiculite, amorphous silicate, or perhaps a clean absorbent clay from the school shop. These materials have a very long shelf life. (Some organizations use buckets of sand mixed with 10 percent soda ash for absorbing acid and alkali spills. And some use activated charcoal for picking up flammable solvents since it has a high affinity for some common organic vapors. Neither is recommended in school labs.) Of these, the amorphous silicate is perhaps the most efficient, absorbing a relatively large volume of liquid in a short period of time (about 30 seconds for 98 percent recovery). It can be purchased in porous bags of various sizes to match the volume of the spill. The spill-saturated bags can be quickly removed from the lab if safety so dictates.[6,7]

Neutralizing acid or alkali spills directly on the floor is not recommended. Neutralization is an exothermic reaction and can cause boiling or spattering. Furthermore, the resulting mess may be difficult to clean up (and may not be truly neutral).

The use of paper towels (cellulose) to pick up spills of strong oxidizing agents such as concentrated sulfuric acid can be dangerous.

The *MCB Chemical Reference Manual*[4] contains information on handling spills. It is recommended that a safety handbook of this type be on your lab shelf because it explains what to do in case of spills of the major classes of compounds.

---

6. "Some Common Misconceptions in Handling Laboratory Spills," F. W. Michelotti and J. W. Seidenberger, *American Laboratory* 11 (November 1979): 77.

7. J. Jensen, "A Spill Absorbing System for the Laboratory," *American Laboratory* 12 (January 1980): 72-78.

Safety in Chemistry Settings 67

One special note about mercury: mercury spills are common and sulfur is often mentioned as an effective agent for clean-up. Actually it is better to improvise a "vacuum cleaner" made of a filter flask connected to an aspirator or a rubber sponge to absorb mercury. (If mercury is spilled in an oven or on a hot surface, evacuate the room promptly.)

IOWA DEPARTMENT OF ENVIRONMENTAL QUALITY REGULATIONS

The information provided in this section is based on Iowa Department of Environmental Quality (DEQ) regulations. These regulations are generally indicative of state regulations throughout the United States. It is recommended that the reader contact his/her environmental quality office for specific state regulations.

A few pertinent regulations regarding disposal are listed below.

Waste disposed into streams and rivers: Section 455B.48 of the Iowa Code prohibits disposal of pollutants into waters of the state unless you have a DEQ permit (chemicals are pollutants by the DEQ's definition). So most schools in the state cannot discharge chemicals directly into the waters of the state. Any school system that does have their own disposal system participates in the DEQ permit program and must follow Chapter 17 (Effluent and Pretreatment Standards; Other Effluent Limitations or Prohibitions) and Chapter 19 (Waste Water Construction and Operation Permits) of the DEQ standards.

Waste disposal into the sewer system: DEQ rule 17.1(7) prohibits the disposal of waste in such volume or quantity that it would interfere with the operation or performance of the treatment system. Pretreatment may be necessary with DEQ rule 17.6(4)(c) being the most pertinent. It states that no waste can be dumped that would intermittently change the pH of the raw sewage reaching the treatment plant by more than 0.5 pH units or cause the pH of the water reaching the plant to be less than 6.0 or greater than 9.0.

Disposal on or in the land: Open dumping of chemicals is prohibited by Section 455B.82 of the Iowa Code and DEQ rule 26.2. Dumping of chemicals in a landfill may be acceptable if the wastes are not toxic or hazardous. The DEQ definition of "toxic and hazardous" is waste materials, including but not limited to poisons, pesticides, herbicides, acids, caustics, pathological wastes, flammable or explosive materials, and similar harmful wastes which require special handling. They can be dumped at a sanitary landfill only with prior written approval from the DEQ.

CHEMTREC--1-800-424-9300

If there is a *major chemical transportation accident* in your locality, assistance is available from CHEMTREC. CHEMTREC is the manufacturing Chemists Association's name for its Chemical Transportation Emergency Center. The center provides immediate information regarding procedures in case of spills, leaks, fires, or exposures.

When calling the toll-free number you will be asked to identify the accident location, the name of the chemical product(s), the nature and extent of the accident, the shipment source, the names of the company that made shipment, the carrier and consignee, whether there are injuries and any local conditions that may affect the hazard. The CHEMTREC Communicator will provide the caller with pre-established information on file such as the kind of hazards to be expected from the product involved, and what to do in case of spills, leaks, fire, or exposure. The communicator then will relay the details of the accident immediately by phone to the shipper, who becomes responsible for any future action in regard to the emergency.

REFERENCES

Craig, Ralph N., "Disposing of Harmful Lab Wastes," *Chemists-Analyst* 68, no. 2 (October 1979): 3-4.

Fawcett, H. H., and Wood, W. S., *The Care, Handling and Disposal of Dangerous Chemicals*. New York: John Wiley and Sons, 1965.

Laboratory Waste Disposal Manual. Washington, D.C.: Manufacturing Chemists Association, 1975.

Mento, Mary Ann, "Chemical Disposal for a High School Chemistry Laboratory," *Science Teacher* 24 (January 1973): 30-32.

National Fire Codes, Vol. 1, Flammable Liquids. Boston, Mass.: National Fire Protection Association, 1976.

Standard on Fire Protection for Laboratories Using Chemicals, No. 45. Boston, Mass.: National Fire Protection Association, 1976.

Standards for Protection against Radiation, Title 10. Nuclear Regulatory Commission, Code of Federal Regulations, Washington, D.C.

Steere, N. V., ed. *CRC Handbook of Laboratory Safety*, 2nd ed. Cleveland, Ohio: Chemical Rubber Co., 1971.

L A B E L I N G

Proper labeling is fundamental to a safe, effective lab operation. It has two principal functions, adequate identification and precautionary information for safe handling.

Purchased reagents have a label that specifies the chemical or common name, the manufacturer, and the lot number. In addition, appropriate information on flash points, hazards, or other precautionary data are usually shown. When the reagent reaches the school storage area, the receiving date should be marked on the label.

Materials that are made in the lab should also be adequately labeled if they are to be kept, including the name of the preparer. Any substance that is not properly labeled should be disposed of immediately.[1]

Do not assume that all manufacturers label chemical reagents properly. Furthermore, don't assume that because one hazard is listed that a reagent possesses only that one hazard. A "hierarchy of hazards" may have been followed in which only the primary hazard is listed. Thus, perchloric acid may be an oxidizer, and toxic and corrosive but only labeled as an oxidizer.[2]

The NIOSH Criteria Document, "An Identification System for Occupationally Hazardous Materials," and NFPA (704-1975) recommend that the label contain the following:

A. The trade name or chemical name of the product.
B. A hazard symbol consisting of three rectangles containing terse indications of relative health hazard, fire hazard, and reactivity hazard.
C. Appropriate statements on the nature of the hazard.
D. Appropriate action statements.
E. Emergency action and first aid statements.
F. Clean-up and disposal statements where appropriate.

---

1. *Safety in the School Science Laboratory*, U.S. Department of Health, Education and Welfare (NIOSH), Cincinnati, Ohio 45226, 1977, pp. 8-3 to 8-7.
2. *Hazardous Materials Safety Seminar*, National Hazards Control Institute, Starson Corporation, Stanton, N.J. 08885, 1978, pp. 2-7.

Safety in Chemistry Settings   69

Figure 5.2 shows the basic concept of the hazard diagram and Figure 5.3 explains the number scale the NFPA recommends in labeling. The diagram is divided into four segments. The top segment indicates the flammability hazard. The left segment indicates the health hazard and the right segment indicates the reactivity. The bottom segment is used to identify any special characteristics that the handler should be aware of. For example, a W with a line through the middle indicates that the chemical should not come in contact with water.[3]

Each number scale ranges from 0 (no hazard) to 4 (extreme hazard). This is superior to the older method of including vague statements on the label such as "avoid skin contact" or "flammable."

An example of the system (for ethyl ether) is shown in Figures 5.2 and 5.3. The health hazard is 2, the fire hazard 4, and the reactivity 1, indicating a moderate health hazard, extreme fire hazard, and low reactivity.

The health hazard designation refers only to the immediate acute effects of exposure to the chemical. Long-term effects are not listed.

All chemicals with a hazard rating of 2 or greater in any category should be stored and handled with due caution. Some safety personnel recommend that secondary students not be allowed to handle stock containers that have any hazard rating of 2 or greater unless the teacher is present.

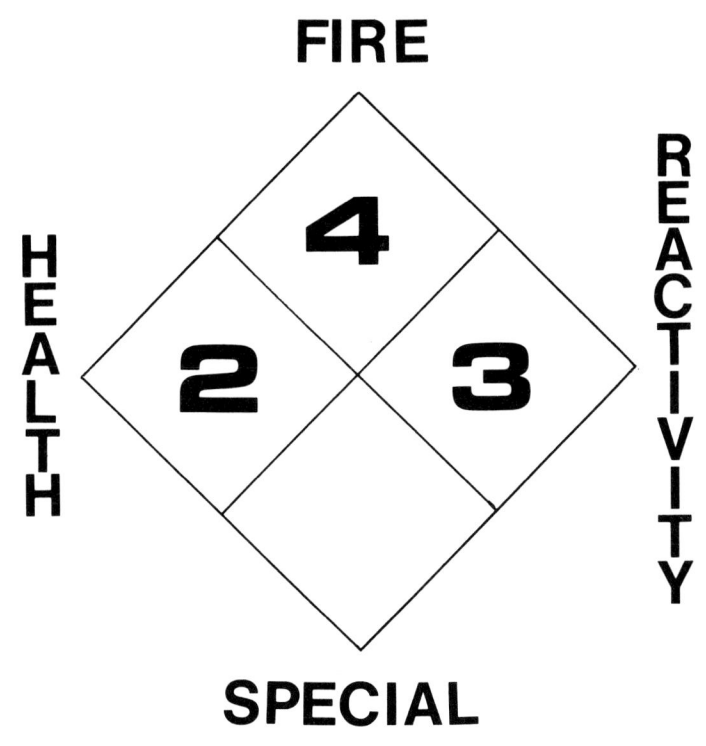

Fig. 5.2. Hazard Diagram.

---

3. *Safety in the School Science Laboratory*, pp. 8-7 to 8-14.

Identification of Health Hazard Color Code: BLUE		Identification of Flammability Color Code: RED		Identification of Reactivity (Stability) Color Code: YELLOW	
Type of Possible Injury		Susceptibility of Materials to Burning		Susceptibility to Release of Energy	
Signal		Signal		Signal	
4	Materials which on very short exposure could cause death or major residual injury even though prompt medical treatment were given.	4	Materials which will rapidly or completely vaporize at atmospheric pressure and normal ambient temperature, or which are readily dispersed in air and which will burn readily.	4	Materials which in themselves are readily capable of detonation or of explosive decomposition or reaction at normal temperatures and pressures.
3	Materials which on short exposure could cause serious temporary or residual injury even though prompt medical treatment were given.	3	Liquids and solids that can be ignited under almost all ambient temperature conditions.	3	Materials which in themselves are capable of detonation or explosive reaction but require a strong initiating source or which must be heated under confinement before initiation or which react explosively with water.
2	Materials which on intense or continued exposure could cause temporary incapacitation or possible residual injury unless prompt medical treatment is given.	2	Materials that must be moderately heated or exposed to relatively high ambient temperatures before ignition can occur.	2	Materials which in themselves are normally unstable and readily undergo violent chemical change but do not detonate. Also materials which may react violently with water or which may form potentially explosive mixtures with water.
1	Materials which on exposure would cause irritation but only minor residual injury even if no treatment is given.	1	Materials that must be preheated before ignition can occur.	1	Materials which in themselves are normally stable, but which can become unstable at elevated temperatures and pressures or which may react with water with some release of energy but not violently.
0	Materials which on exposure under fire conditions would offer no hazard beyond that of ordinary combustible material.	0	Materials that will not burn.	0	Materials which in themselves are normally stable, even under fire exposure conditions, and which are not reactive with water.

Fig. 5.3. NFPA Hazard Coding System. (Reprinted by permission from NFPA 704, *Recommended System for the Identification of Fire Hazards of Materials*, Copyright 1975, National Fire Protection Association, Boston, Mass. 02210.)

Safety in Chemistry Settings

It is recommended that students be taught the principles of this labeling system to increase their awareness of hazards in the lab. Since the DOT requires this labeling system for shipping of chemicals the student may have an opportunity to make use of the information outside the school experience.

It is a good idea to label the storage area where *corrosive* or *hazardous* materials are kept to help insure that the compounds are returned to the appropriate place.

Labels for Carcinogens. It has been recommended in other sections of this manual that carcinogens not be kept in school science labs. If a valid reason for keeping a carcinogen does exist, it must be identified by the full chemical name and the Chemical Abstracts Registry number. Containers must have the warning "CANCER-SUSPECT AGENT" directly under or next to the content identification. If the carcinogen is associated with any other health hazards they must also be indicated on the label.[4]

REFERENCES

Compilation of Labeling Laws and Regulations for Hazardous Substances, Chemical Specialties Manufacturers Association, Inc., 50 East 41st Street, New York, N.Y.

Federal Hazardous Substances Act Regulations, Code of Federal Regulations 16, part 1500, U.S. Government Printing Office, Washington, D.C. 20402, 1974.

Guide to Precautionary Labeling of Hazardous Chemicals, Manual L-1, 7th ed., Washington, D.C.: Manufacturing Chemists Association, 1970.

An Identification System for Occupationally Hazardous Materials, A Recommended Standard, National Institute for Occupational Safety and Health, HEW Publication No. (NIOSH) 75-126, U.S. Government Printing Office, Washington, D.C. 20402, 1974.

Manufacturing Chemists Association, *Guide for Safety in the Chemical Laboratory*. New York: Van Nostrand Reinhold, 1972.

Recommended System for the Identification of Fire Hazards of Materials, NFPA 704-1975, National Fire Protection Association, 470 Atlantic Avenue, Boston, Mass. 02210.

Steere, N.V., "Containers in Labeling," in *CRC Handbook of Laboratory Safety*, 2nd ed. Cleveland, Ohio: Chemical Rubber Co., 1971.

Warning Labels, Manual L-1, 1970, 7th ed. Washington, D.C.: Manufacturing Chemists Association.

RADIOACTIVE MATERIALS

REGULATION

The Iowa Department of Environmental Quality is the state agency designated to establish policy for transportation, storage, handling, and disposal of radioactive materials for the purpose of protecting the public health and safety (Chapter 455B.86 of the Iowa Code). But the *Rules and Regulations of the Nuclear Regulatory Commission* (NRC) are in force in Iowa.

Radioactive sources are usually supplied to schools under a general license by various manufacturers. The activity level of these isotopes is low and they are considered safe to handle with minimal precautions. Table 5.9

---

4. *Carcinogens--Regulation and Control*, U.S. Department of HEW, NIOSH, Publication No. (NIOSH) 77-205, Cincinnati, Ohio 45226, 1977, p. 30.

Table 5.9. SOME READILY AVAILABLE RADIOISOTOPES

Radioisotope and Type of Radiation	License Exempt Quantity (microcuries) μC	Half-life
Calcium--45 (β)	10	165 days
Carbon--14 (β)	100	5570 years
Cerium--144 (β,γ)	1	284 days
Cesium--137 (β,γ)	10	30 years
Chromium--51 (γ)	1000	27 days
Cobalt--60 (β,γ)	1	5.3 years
Iodine--131 (β,γ)	1	8.1 days
Phosphorus--32 (β)	10	14.3 days
Polonium--210 (γ)	0.1	138 days
Strontium--90 (β)	0.1	29 years
Thallium--204 (β)	10	3.8 years

lists some readily available radioisotopes and their maximum allowable activity level.[1]

Film badges and dosimeters are not necessary when handling license exempt quantities of radioisotopes.[2] Teachers should confine their use to the license exempt quantities taken from Part 30, Section 71, Schedule B of the *Rules and Regulations of the Nuclear Regulatory Commission*.

HANDLING RADIOISOTOPES

The sealed radioactive sources are very convenient and should be chosen over the liquid solution if possible.

When working with radioactive liquids always work over a tray lined with absorbent paper to catch spills. It is also recommended that the counter top be lined with a plastic-backed absorbent paper or a layer of paper towel with wax paper underneath.

Spills should be wiped up immediately. Wear disposable gloves when handling liquid radioactive sources. Monitor the gloves for contamination frequently. Never pipette radioactive liquids by mouth.

Use of radioisotopes with animal experiments is not recommended, but if done, take the following precautions:

1. Anesthetize animals adequately before injecting them.
2. Use disposable syringes.
3. Keep animals that have been injected or fed radioactive materials securely locked in labeled cages with absorbent paper below.

Radioactive wastes need to be placed in a container specified for that purpose.

Keep all acids away from I^{131} (supplied as NaI) and C^{14} (supplied in

---

1. *Rules and Regulations of the Nuclear Regulatory Commission*, Part 30, Section 71, Schedule B.
2. John W. Sulcoski, "The Safe Use of Radioisotopes," *Journal for Students and Teachers* 5, issue 30 (May 5, 1965).

Safety in Chemistry Settings

carbonate form) to avoid generation of radioactive gases. If there is a chance of radioactive gas generation, work in the fume hood. Do not leave the work area until hands, clothing, and shoes have been monitored for radioactivity.

Be sure you have monitors that will detect the type of radiation you are handling. Remember that a good gamma ray detector may be very poor at detecting beta rays.

NRC approved labels should be used on all glassware containing radioactive materials. The label should indicate which radioisotope is inside, the activity level and the date.

Use forceps or tongs to handle radioactive materials.

Keep a record of radioisotope purchases, date of receipt, quantities used, user names, and spills.

Keep equipment, glassware, and the materials that are not involved in the immediate operation away from the area.

RADIOISOTOPE WASTE DISPOSAL

For license exempt quantities of radioactive materials (in the microcurie range of activity), follow these simple rules:

1. Dilute solutions with large amounts of water and wash down the drain.
2. Incinerate solid materials in a properly ventilated area.

Do not allow radioactive wastes to accumulate. Get rid of them on a regular basis.

If the school has its own septic tank or sewage disposal system, incineration would be a better choice.

Working with higher activities than those allowed in schools under the license exempt provision involves more sophisticated disposal techniques such as hiring a licensed contractor to remove wastes. Follow NRC guidelines.

RADIOISOTOPE STORAGE

Radioactive materials should be stored in a locked storage area that is labeled as such. Thick shielding using lead bricks is not necessary for license exempt quantities.

The relative penetrating powers of the alpha, beta, and gamma are 1, 100, and 10,000 respectively. Students should be familiar with this generalization. The skin will stop normal alpha radiation. However, ingestion is the most hazardous potential in the school lab, and once inside the body, even alpha emitters are highly radiotoxic. This is particularly true of isotopes that tend to concentrate in one organ of the body. If ingestion does occur, contact a doctor immediately.

RADIATION DECONTAMINATION PROCEDURES

When finishing an experiment involving radioactive materials, a survey of the personnel, labware, and work area should be completed using a meter that is sensitive to the type of radiation being emitted. If the radiation level is double the normal background rate follow the appropriate procedure below:[3]

Glassware and sinks--Scrub with a strong lab detergent such as Alconox if

---

3. *Science Classroom and Laboratory Procedures*, Waterloo Community Schools, Waterloo, Iowa, 1968, p. 41.

available. Household detergents can be used also. Use plenty of water for rinsing. If you cannot decontaminate the glassware with even the strongest detergent, throw it away.

Clothing--Use detergent and water. Rinse well.

Skin--Gently rub contaminated area with a soft brush containing soap and water.

Concrete, painted surfaces, metal surfaces--Wash with detergents. For stubborn contaminates, try dilute HCl unless it will generate radioactive gas (C^{14} or I^{131}).

FREQUENTLY USED TERMS

Roentgen (r)--a unit for expressing exposure from X rays or gamma rays in terms of the ionization produced in air. It is the amount of radiant energy that produces 1.61×10^{12} ion pairs in one gram-centimeter of air.

Rad--a unit of "absorbed dose" equivalent to 100 ergs/gram (regardless of medium or ionizing radiation type).

Roentgen equivalent man (rem)--a unit used to express the estimated equivalent of any type of radiation that would produce the same biological end point as one rad delivered by X-rays or gamma rays.

Curie (Ci)--the most common unit used to express the radioactivity of a material. It is the amount of any radionuclide (or combination of radionuclide) in which there are 3.7×10^{10} disintegrations per second. Related designations include:

$$1 \text{ millicurie (mCi)} = 10^{-3} \text{ Ci}$$
$$1 \text{ microcurie (}\mu\text{Ci)} = 10^{-6} \text{ Ci}$$
$$1 \text{ nanocurie (nCi)} = 10^{-9} \text{ Ci}$$
$$1 \text{ picocurie (pCi)} = 10^{-12} \text{ Ci}$$

Half-life--the time required for a radioisotope to lose half of its activity by decay.

Biological half-life--the time it takes the body to eliminate one-half of an administered dose of any substance by the regular elimination routes.

REFERENCES

Boursnell, J., *Safety Techniques for Radioactive Tracers*. Cambridge, England: Cambridge University Press, 1958.

Control and Removal of Radioactive Contamination in Laboratories, National Bureau of Standards Handbook, No. 48, U.S. Government Printing Office, Washington, D.C. 20402.

Fire Protection Handbook, Chapter 13, "Nuclear Reaction," Radiation Machines and Facilities Handling Radioactive Materials, NFPA, 470 Atlantic Avenue, Boston, Mass. 02210.

Glossary of Terms in Nuclear Science and Technology. ANSI NI.1, New York: American National Standards Institute, 1957.

The National Institute of Health Radiation Safety Guide, U.S. Department of Health, Education and Welfare, DHEW Publication No. (NIH) 73-18, U.S. Government Printing Office, Washington, D.C. 20402, GPO Bookstore Stock Number 1740-00351.

Radiological Monitoring Methods and Instruments, National Bureau of Standards Handbook No. 51, U.S. Government Printing Office, Washington, D.C. 20402.

Standards for Protection against Radiation, Title 10, Code of Federal Regulations, U.S. Nuclear Regulatory Commission, 1717 H Street, N.W., Washington, D.C.

Shapiro, J., *Radiation Protection*. Cambridge, Mass.: Harvard University Press, 1972.

Steere, N. V., *CRC Handbook of Laboratory Safety*, 2nd ed. Cleveland, Ohio: Chemical Rubber Company, 1971.

OTHER RADIATION SOURCES

X RAYS

Any device that uses electron beams, such as Crooke's tubes (cathode ray tubes) or several of its variations, can produce X rays. In fact, any instrument in the lab that produces high energy electrons that strike a metallic target in a vacuum or low pressure tube should be held suspect.[1]

Avoid exposing anyone to X rays unnecessarily by:

1. Performing Crooke's tube experiment by demonstration only and for a brief time.
2. Operating the tubes at the lowest working voltage.
3. Keeping students several feet away from the tube during operation.

LASERS

Even low output lasers can be a hazard because long-term eye exposure can cause retinal damage. The primary damage mechanism is simple heating caused by the absorption of concentrated light energy. Be particularly cautious of lasers purchased prior to 1976; they may not be labeled with output power or may vary widely from the stated output.[2]

There is no skin hazard for the typical helium-neon laser.

Most lasers found in school labs are rated at five milliwatts or less and a few simple precautions should be followed to avoid risk:

1. Never view a laser beam in such a way that the beam travels directly into the eye.
2. Avoid aiming the beam at any reflective surfaces that would "bounce the beam into the eye.
3. Block the beam whenever possible.
4. When not in use, keep the laser secure and turned off. Don't leave it where it could be played with.
5. Use the minimum output power necessary to accomplish the goal of the experiment.
6. Keep the area well lighted so that the pupils of everyone's eyes are relatively contracted.
7. When possible, bench mount the laser and test the experiment without having students present.

---

1. *STOP--Safety First in Science Teaching*, Division of Science, North Carolina Department of Public Instruction, 1977, pp. 28-29.
2. *STOP*, p. 32.

REFERENCES

"Lasers and Masers--Control of Health Hazards," *Journal of Chemical Education* (March 1965).

Laser Fundamentals and Experiments, Office of Information, Bureau of Radiological Health, Food and Drug Administration, 5600 Fishers Lane, Rockville, Md. 20857.

"Safe Use of Lasers," Z-136.1, 1976, American National Safety Institute, 1430 Broadway, New York, N.Y. 10018.

"Safety in Classroom Laser Use," Office of Information, Bureau of Radiological Health, Food and Drug Administration, 5600 Fishers Lane, Rockville, Md. 20857.

Sliney, D. H., and Palmisano, W. A., "The Evaluation of Laser Hazards," *American Industrial Hygiene Association Journal* 29 (1968):425.

6

Safety in Physics Settings

Safety in the physics laboratory and classroom requires continued attention that is primarily the responsibility of the instructor. It is the instructor's duty to conduct the classes in a manner that reflects concern for the students' welfare and will develop in the students a safety conscious attitude and behavior. Much of the equipment used in the teaching of physics, if not properly operated, could present problems related to safety.

CUTS OR PUNCTURES
Many kinds of such motorized equipment as vacuum pumps, blowers, and power tools have belts and pulleys that must have guards to reduce the possibility of clothing or hands entanglement that could result in injuries such as cuts or punctures. Students need instruction in cutting tools (for example, razor blades and knives) and the proper use of such items as glass tubing and glass plates. Students should wear gloves to reduce potential injury.

BURNS
Many items such as heat sources present safety problems related to burns. Heat lamps, bunsen burners, alcohol lamps, and candles, all of which have a visible indication of heat to indicate danger, have the added hazard of more easily igniting materials.
"Invisible" sources, such as hot plates, ovens, and hot glass or metal, need special attention. Specific procedures for use, such as (1) specific locations away from careless contact, (2) warning signs or barriers when in operation, and (3) specific operating instructions are the best safety techniques for all such heat items. For additional safety (1) restrict loose clothing, (2) restrict loose hair, and (3) use heat resistant gloves and other special handling tools.

EXPLOSION/IMPLOSION
Equipment that operates under pressure should be frequently checked for proper operating instructions and proper functioning of safety valves or automatic cut-off switches. These precautions will reduce the possibility of an explosion. Equipment that operates or is in a state of reduced pressure (vacuum), such as dewar flasks, vacuum bell jars, and cathode ray tubes, should be frequently checked for weaknesses that might cause implosion. Shields should be used wherever possible. Refrigerators and other storage devices that are used for potentially explosive materials should be specifically designed, or

adapted, for such use. Household refrigerators are not ordinarily explosion-proof.

ELECTRICAL

Electrical equipment is a common potential source of serious injuries. Line voltages (110 volts AC) and equipment connected to the line must be insulated and/or shielded to prevent contact with current. Such equipment should also be grounded. Equipment or outlets in the laboratory should have ground-fault interrupters that would automatically cut off current faster than fuses or circuit breakers, thus reducing possibility for severe injury from contact with electrical current.

High voltage equipment should be provided with special interlock or key-operated switches so they can be activated only under specific, safe, and supervised conditions.

High voltage-low amperage electricity, such as Van de Graff or other static charged equipment, should be shielded from human contact. Each individual person reacts in a different way to electrical current or shock. It is NEVER safe for an individual to receive electrical current or shock.

Students should be instructed never to "short circuit" dry cells or storage batteries since high temperatures developed in connecting wires can cause burns.

When charging any type of capacitor, insure that the rated capacity is not exceeded. In a circuit that contains a fluctuating DC voltage electrolytic type capacitors which are polarized (±) can be used. When connecting this type of capacitor to points in a circuit use extreme caution to make sure that the proper polarity rules are being followed (negative to negative and positive to positive). If the capacitor is connected backwards an explosion could result.

Electrical appliances and equipment should show listing and approval by Underwriter's Laboratories, Inc., or another nationally recognized testing laboratory.

CRYOGENIC

When there is a possibility of personal contact with a cryogenic fluid, full face protection, an impervious apron or coat, cuffless trousers and high-topped shoes should be worn. Watches, rings, bracelets should not be permitted when working with cryogenic fluids.

RADIATION

Radiation from different kinds of sources requires special safety precautions. When using lasers in the classroom or laboratory, overexposure of the eye and skin is of primary concern, but the hazard of electrical shock exists also. The lasers found in most secondary schools are helium-neon lasers of less than 5 milliwatt power and overexposure is much less of a safety problem. However, it is wise to have the student follow the general safety rules for lasers. The rules given below are recommended by the U.S. Department of Health, Education and Welfare.

1. Work Area Controls
   a. The laser should be used away from areas where the uninformed and curious would be attracted by its operation.
   b. The illumination in the area should be as bright as possible in order

Safety in Physics Settings 79

      to constrict the pupils of the observers.
- c. The laser should be set up so that the beam path is not at normal eye level, below 3 feet or above 6½ feet.
- d. Shields should be used to prevent both strong reflections and the direct beam from going beyond the area needed for the demonstration or experiment.
- e. The target of the beam should be a diffuse, absorbing material to prevent reflection.
- f. Remove all watches and rings before changing or altering the experimental setup. Shiny jewelry could cause hazardous reflection.
- g. All exposed wiring and glass on the laser should be covered with a shield to prevent shock and contain any explosions of the laser materials. All nonenergized parts of the equipment should be grounded.
- h. Signs indicating the laser is in operation and that it may be hazardous should be placed in conspicuous locations both inside and outside the work area and on doors giving access to the area.
- i. Whenever possible, the door(s) should be locked to keep out unwanted onlookers during laser use.
- j. The laser should never be left unattended.
- k. Good housekeeping should be practiced to insure that no device, tool, or other reflective material is left in the path of the beam.
- l. A detailed operating procedure should be outlined beforehand for use during laser operation.
- m. Whenever a laser is operated outside the visible range (such as a CO_2 laser), some warning device must be installed to indicate its operation.
- n. A key switch to lock the high voltage supply should be installed.

2. Personnel Control
   - a. Avoid looking into the primary beam *at all times*.
   - b. Do not aim the laser with the eye; direct reflection could cause eye damage.
   - c. Do not look at reflections of the beam; these, too, could cause retinal burns.
   - d. Avoid looking at the pump source at all times.
   - e. Clear all personnel from the anticipated path of the beam.
   - f. Do not depend on sunglasses to protect the eyes. If laser safety goggles are used, be certain they are designed to be used with the laser being used.
   - g. Report any afterimage to a doctor, preferably an ophthalmologist who has had experience with retinal burns, as damage may have occurred.
   - h. Be very cautious around lasers which operate in invisible light frequencies.
   - i. Before operation, warn all personnel and visitors of the potential hazard. Remind them that they have only one set of eyes.

    Ultraviolet lamps should always be properly shielded and eye protection worn if the source is exposed. No source should be viewed directly. It should be noted that an electric arc produces a large amount of ultraviolet and is dangerous as an open source.

    Equipment producing electron beams such as cathode ray tubes and Crooke's tubes can produce X rays when used at high voltages. When demonstrating with Crooke's tube have the apparatus on for only short periods of time and have the students several feet from the source.

    Apparatus used to look at objects, either in the sky or on the ground, such as binoculars, telescopes, sextants, octants, levels, and transits must be carefully supervised and proper procedures carefully instructed and observed. Some items, such as sextants, will have sun shields, which should be used when sun ray incidence is possible. Individuals should never be permitted to look directly through any apparatus at the sun, even during an eclipse.

Sources of radioactive radiation should only be used under controlled conditions with personal protection. Please refer to page   of the Chemistry section on radioactive materials.

VISIBILITY

Many accidents occur through lack of visibility as well as too bright a light.

Fifty foot-candles of illumination is recommended as minimum levels in any area. Painted stripes, indicating limits of door swings or apparatus operation, provide possible protection from injury due to invisible or intermittent operation (swinging doors). Fluorescent stripes or paint on step edges, corners or posts will provide visibility during power failure or emergency exits during smoky fires.

Invisible hazards, such as radiation, biological, chemical, electrical, and heat should all be signaled or located by use of adequate procedures of signs and warning symbols.

IMPACT

Falling items in the laboratory can cause severe injury to head, body, appendages, particularly feet. Tall racks, if used at all, should be well anchored to prevent shaking or toppling. Heavy and/or unstable objects should be placed on lower shelves.

REFERENCES

Armitage, Phillip, and Fasemore, Johnson, *Laboratory Safety: A Science Teacher's Source Book*. London: Heinemann Educational Books, 1977.

Everett, K., and Jenkins, E. W., *A Safety Handbook for Science Teachers*. London: John Murray, 1977.

Martin, Alan, and Harbison, Samuel A., *An Introduction to Radiation Protection*. London: Chapman and Hall, 1972.

Radiation and Protection in Educational Institutions. Washington, D.C.: National Council on Radiation Protection and Measurements, 1966.

Safety in Classroom Laser Use. Rockville, Md.: U.S. Department of Health, Education and Welfare, Bureau of Radiological Health, May 1970.

Safety in the Secondary Science Classroom. Washington, D.C.: National Science Teachers Association, 1978.

Steere, N. V. *CRC Handbook of Laboratory Safety*, 2nd ed. Cleveland, Ohio: Chemical Rubber Company, 1971.

7

Field Activities

The section on Safety in Biology Settings contained a number of references to field activities. However, a separate discussion is included here since field trips are generally regarded as especially important parts of work in biology and earth science.

No safety checklist can take the place of the foresight and carefully laid plans of a reasonable and prudent teacher in the planning of field trips. While it is impossible to anticipate every conceivable hazard that might present itself on any given field trip, some general guidelines can cover the most important precautions to consider. These guidelines will suggest the more important areas to consider in providing for the maximum safety of the students involved. (See also Appendix A, p. 112.)

CONSENT, PERMISSION, AND APPROVAL

It is important that the appropriate individuals be contacted for permission to take students into the field well in advance of the activity. The question may arise as to what constitutes a field trip or field activity for which special permission and approval is needed. Although local school policy may require some form of information being given, or even formal approval, to take any group out of the scheduled time or room, this section is concerned primarily with trips beyond the schoolyard. Most local schools will have definite, written policies concerning the steps to take in planning and securing approval for a field trip. Teachers, of course, should be thoroughly familiar with these policies.

Three groups of people should be considered in securing permission and approval. These are parents, school authorities, and those in charge of the site being used or visited. If field activities are common in a particular class or a particular school, forms may be designed for use in systematizing the process.

Parents should be expected to give their written consent for a student to be away from the schoolgrounds for special activities. This is particularly important when the student is separated to a significant degree, not only as to site, but also as to time schedule. Weekends or evenings are examples of such a significant degree. When parents give their consents, the following should be obtained: names and telephone numbers where a parent or the parents can be reached in the case of an emergency; injury release form; and instructions concerning any student requiring special medical consideration. Some of this may be kept in permanent records and checked periodically. This is especially appropriate when frequent field work is done. The injury release form should be considered separately for each trip or activity. It may not be necessary in every case, but its omission should be rare. Medical considerations should include allergies, chronic conditions, current health conditions, and medication being used.

Of course, school authorities should be contacted and the necessary forms should be on file in school offices. Again, it should be noted that each local school will probably have its own school policies and procedures governing field trips. If a school does not have such policies, science teachers should encourage the adoption of such policies and should exercise leadership in obtaining them.

It is important that on-site authorities be contacted early in the planning for permission. On-site agents may be owners of areas and have very little contact with the activity. Others will be tour leaders or formal consultants. These leaders or consultants may have a great deal of involvement in the project. They may have very valuable recommendations and information to be used in preparation for the trip.

In some cases, it may be necessary to have legal advice as to liabilities involved or how protection of the teacher against liability may be threatened. Insurance forms may need to be filled out. Insurance of individuals should be carefully considered in the case of each trip involving transportation and more than ordinary hazards to be encountered during the project.

SUPERVISION

Although the regular teacher will not always have the sole responsibility for immediate supervision of each individual, it is that teacher's responsibility to provide for adequate supervision. There should be at least one supervisor or chaperone for each eight to ten students. The exact degree of the supervision or chaperonage expected should be discussed carefully with the class and with the extra persons.

A schedule should be established. The class will usually be involved in the development of the schedule. Of course, it will be recognized that revisions in the schedule may be necessary, but it should be established with care and modified only with legitimate bases.

A schedule should include a number of mustering times and places when and where the roll may be taken, revisions in plans noted, and information not available before can be shared. These round-up sites should be selected in such a way that they can be readily located since students who may become separated from the group during the session may be reunited with the class at these places.

Extra supervisors should become fully familiar with the members in the subsets with which they are working. These supervisors may be chosen for a number of functions. In some instances, they may serve as expert consultants and advisers for the projects. In other instances, they may be chosen for their capabilities in working with young people and in motivating serious work. They should meet the class before the trip and have some opportunity to work with the ones with whom they are assigned in a more stable setting than in the field. The teacher will also have spent some time with the extra people giving detailed information about the trip and describing how chaperones and supervisors can help in the activity. Careful attention should be given to sharing the educational objectives of the activity and it would be hoped that these assistants would be selected for the educational contributions they could make, as well as for their ability to demand respect and maintain order. The teacher should personally select the aides. Requests should not be made of a parents group asking it to select the individuals. Chaperones with first-aid competencies are usually to be desired.

Field Activities

TRANSPORTATION

School transportation should be used if at all possible. School vehicles are designed for safety of students and school drivers are accustomed to behavior patterns of students.

Explicit rules of behavior should be discussed with students before the activity. As with all laboratory activity, work in the field demands serious behavior. Careless and foolish acts cannot be tolerated. Pre-trip planning discussions should emphasize this and should point out that even single digressions cannot be forgiven. Serious behavior is especially important during transportation periods.

Not all students in a class will have had experience in the form of transportation being used, a school bus, for example; therefore, instruction should be given at the beginning of the trip about passenger safety, including emergency evacuation procedures.

Backup transportation in the event of an emergency should be considered. It may not be necessary to have it standing by in readiness, but the project leader should be knowledgeable as to steps to take in securing replacement transportation.

PRE-TRIP SITE VISIT

The site should be visited by the teacher before the field trip. In many instances, the other chaperones and supervisors should accompany the teacher on the pre-trip visit. A number of specifics of the plan may be established during such a visit.

A number of specific characteristics of the area should be looked for during the early visit. Sources of potential dangers should be carefully noted, such as deep water, poor trails for walking, and falling rocks. Hazardous trails should be mapped with hazards targeted. These hazards might include barbed wires, steep gradients, rough terrain, and slippery pathways. Poisonous plants and irritants should be located. Safety measures to use with animals in the laboratory are discussed in the section on biological safety. Similar measures should be followed in interacting with animals in the field. Feral animals will usually avoid humans in the field, especially when the humans are in groups. In the event a wild animal does not try to escape, special care should be exercised, because this lack of fear might be the result of impaired senses or disease conditions, such as hydrophobia. If one is bitten by an animal in the field, the animal should be captured, if possible. This capture should be by the teacher and supervisors using extreme caution. The animal should be observed carefully over a period of two to three weeks. Of course, the wound from the bite should be treated, both in the field and after return to the school. Bravado and showmanship should not be practiced in the field in capturing animals. Nonpoisonous snakes might be picked up rather easily, but their bites can produce unnecessary infections. Sometimes a poisonous snake might be picked up by mistake because quick identification in the field is not always reliable.

COLLECTIONS

Collections made in the field should be held to a minimum for a number of reasons. Clearly understood objectives should be the basis for collecting specimens and returning them to the school. Assignments should be made for collections. Collecting for the sake of collecting should be avoided. The integrity of the ecosystem in a natural environment should be respected. Killing jars, preservatives, and other collecting apparatus and techniques are discussed in the biological section. Endangered species should be avoided, and legal restrictions regarding collections should be carefully checked.

OTHER CONSIDERATIONS

Safety equipment should be provided. This might include such items as life jackets to be used in or near bodies of water, protective eyeglasses when rocks are being struck with rock hammers, and protective helmets where there is danger of falling objects. Special safety equipment may be arranged for at the site itself in the case of certain field trips. Clothes worn on trips should be appropriate for the sites being visited and for the weather.

Both students and supervisors should know individuals and agencies to contact in the event of an emergency. School authorities should be reminded of trips and enough of the details to be alert to help if contacted. In going into certain types of areas, owners or others in the near vicinity should be informed at the time of the trip. This should be done in addition to securing permission prior to the trip. The knowledge of those nearby may be helpful in emergencies.

It may be necessary to change plans for a trip or cancel it due to hazardous conditions created by weather. Regular procedures should be followed in making these changes. Individuals or small groups certainly should not continue on the trip when the regular activity has been cancelled.

Field activities are important aspects of science education. They require careful attention in planning, execution, and followup. The extra effort demanded may reduce a teacher's enthusiasm for involving students in these activities. Systematic planning and experience will counteract the thoughts of reduction, when the results are considered. All field activities should be followed up by restating educational objectives and by reviewing and criticizing the mechanics of the activities themselves. Critiques should give special attention to safety features. The recommendations made following a trip may well avert an accident on the next one.

REFERENCES

Dean, Robert A.; Dean, Melanine; and Motz, LaMoine L., *Safety in the Secondary Science Classroom*. Washington, D.C.: National Science Teachers Association, 1978.

North Carolina Department of Public Instruction, *Safety First in Science Teaching*. New York: Holt, Rinehart, and Winston, 1977.

Sweetser, Evan A., "Field Trips and Teacher Liability," *American Biology Teacher* 36, no. 4 (April 1974): 239-40.

8

Student Research (Projects)

Safety measures to be taken in research projects will be those appropriate for the specific project, the general subject matter area, and the expertise and sophistication of the experimenter. A project undertaken by a junior high school student would require different precautions to be observed and different prohibited practices than one being pursued by a graduate student in biology. Therefore, only a few general statements can be made here. Emphasis, as is true in other sections, will be placed on elementary and secondary activities. Consequently, the term "research" will be somewhat misleading in most instances and could often be replaced by such descriptors as "individual study," "project," or "special investigation." (See also Appendix A, p. 112.)

DESIGN

An important part of any project design will be hazards anticipated and the safety measures to be taken to avoid or deal with these hazards.

The design will also need to provide for adherence to appropriate codes governing use of scientific materials. Humane treatment of animals is one such example. Shielding, necessary in the use of specific radioisotopes, is another example. Guidelines for conducting recombinant DNA experiments as issued by National Institute of Health is still another.

It would be expected that careful attention would be given to the safety factors involved in the project as the design is critiqued and approved. An appropriate reason for a particular design to not be approved would be the lack of attention to safety factors.

SUPERVISION

Any research methodology course, at whatever educational level it is taught, should emphasize safety of the investigator, others in the near vicinity, and the public in general.

Many investigating projects, although worked with at the college or graduate school level, would be inappropriate for an investigator of secondary school age. The project should reflect the skills and knowledge necessary to work in a safe manner. Other projects, although not prohibited entirely, might require much more careful supervision. It is possible that some secondary school projects might be supervised by experts not in the school. A student might work with such types of resource persons as university researchers, physicians, engineers, or industrial researchers.

HOME LABORATORY

The elementary or secondary school cannot take responsibility for laboratories established in the homes of students. Projects expected as a part of regular class requirements or those sponsored by the school should be worked on in school, if at all possible. This will not result in constant supervision, but it will enable more definitive rules to be established for the conduct of the investigation, and the work will be done in an environment where safety measures are expected. Teachers may serve as consultants for establishment of home laboratories, but liability for inadequacies should not rest on a teacher in this consulting role. Some parents may assume responsibility because of their expertise and may be willing to provide the necessary supervision even for a school-sponsored project.

A great number of important scientific breakthroughs have been made only by means of great personal risks. However, this risk is assumed on an individual basis by a mature investigator, and such maturity would be rare until college or graduate school level work is reached.

9

Physical Plant and Facilities

Science rooms are often equipped to provide both classroom and laboratory instruction in a single classroom-laboratory unit. A classroom-laboratory unit is equipped with facilities for instruction in several subject areas. This unit should consist of the following areas:

1. Group instruction
   Classroom area with pupil tables or desk-chair units providing a flat working surface, placed so that pupils may view teacher demonstration unit, chalk board, and projection screen.
2. Laboratory instruction
   Laboratory instruction is provided at selected wall tables or counters equipped with water, gas, and electric utilities; or at 4-pupil peninsular tables; or at island tables in the central room area. When water and gas services are not required, some laboratory instruction may be given at pupil tables in the classroom area.
3. Storage of supplies and equipment
   Storage involves, in addition to perimeter storage in the science room, a special storage area or room. This room may serve more than one science room.
4. Instructor preparation of teaching materials space
   A preparation area or room equipped with counter, sink, and needed utility services and adjacent to storage area is useful.
5. Instructor conference and planning
   This office, conference, and planning area can be incorporated with the preparation area.
6. Individual student study-research
   This unit is an advanced project area with lab counter, sink, service utilities, and reference shelving.

THE SCIENCE PROGRAM
   Traditionally, science may take the form of a generalized science program such as life science, earth science, and physical science, while other specialized science programs usually consist of courses in biology, chemistry, physics, and electronics.

ENROLLMENT IN SCIENCE PROGRAMS
   In the science facility, the needs will depend on the size of the school and the percent of the total enrollment electing a particular science course.

Local science programs should be carefully evaluated and reviewed in planning new or remodeled science facilities.

CLASS SIZE

Science classroom-laboratories are usually equipped to accommodate 24-28 pupils. However, class size is a local board policy determination and may vary considerably, depending on the situation. It would be recommended that no more than two students be placed at a single laboratory station.

DETERMINATION OF TEACHING STATIONS

The following formulas are adequate to determine the number of science teaching stations. For each subject:

1. Number of sections = $\dfrac{\text{Total subject enrollment}}{\text{Class size}}$

2. Number of science classrooms = $\dfrac{\text{Number of sections}}{\text{Periods, less 1, in daily schedule}}$

3. Number of teaching stations =
    (Number of science classrooms) x $\dfrac{\text{Class size}}{\text{Desired group size}}$

SUGGESTED FEATURES OF A CLASSROOM-LABORATORY UNIT

The following suggested floor area (approximate) is for a 28-pupil classroom-laboratory unit:

Classroom-laboratory and ancillary facilities at
    52½ sq ft per pupil. Total: 1470 sq ft

Ancillary facilities would include:

storage	126 sq ft
specialized rooms	126 sq ft
office, conference, and preparation	375 sq ft

with approximate square footage which is included in above total.

SUGGESTED FEATURES OF SPECIALIZED LABORATORY UNIT

Biology

There is considerable similarity in the space and furniture needs of general science and biology courses. Movable equipment varies and the biology laboratory should have auxiliary plant and animal areas. These laboratories may therefore be used for either subject with careful planning or auxiliary facilities. Components unique to biology:

a. Make extensive provision in the biology room for growing tables, aquarium tanks, and animal pens.
b. Provide for microscope storage; or, microscopes may be permanently mounted on work counter along one wall.
c. Consider substage lamps with daylight filters for work with compound microscopes.
d. Provide a utility sink in or near the laboratory for cleaning crocks and

garbage containers; a garbage disposal may be provided in the preparation area and at the demonstration unit sink.
e. A lockable refrigerator is desirable in the biology room or storage room.
f. Provide separate temperature control and humidistat when plant and animal rooms are maintained; separate exhaust ventilation where plant and animal materials are stored; and provision for maintenance of temperature during extended vacation periods when temperature of plants may be lowered.
g. Electric stove, four element, with oven.

Plant Room

An area of about 100 square feet is desirable for special plant projects in biology. This should be a separate room with at least 15 lineal feet of noncorrosive (aluminum, copper, etc.) frame racks 24 inches deep, illuminated by south light or about 1500 footcandles of electric light to produce photosynthesis. There should be a sink with sediment trap, hot and cold water. A floor drain, separate temperature control and humidistat, and water resistant floor materials should also be provided. Plant rooms are sometimes located in bays or attached as "lean to" greenhouses.

Animal Room

This room, when provided, should be at least 100 square feet in area, all surfaces to be "wash down," have a sink with hot and cold water mix with hose connection, floor drain, counter space with storage beneath, humidity and temperature control. Metal frames provide support to various sized animal cages. Electric outlets should be waterproof.

Chemistry

a. The normal ventilation system must be supplemented with a manual control exhaust system which provides a specified number of volume changes of air per minute.
b. The function of a laboratory hood system is to capture, contain, and expel any emissions generated by any operation carried out in the hoods. The operation should be checked frequently and maintained properly to insure adequate operation. A method most commonly used to evaluate the performance of a hood is the face velocity. The face velocity of a hood is the average velocity of the air in feet per minute (fpm) in a direction perpendicular to the plane of the hood opening. It is recommended that for routine laboratory activities requiring a hood, the face velocity be at least 100 fpm. For activities involving potentially hazardous substances (noxious fumes, volatile chemicals, and the like) it is recommended that the face velocity of the hood be increased to between 125 and 200 fpm.
c. The fume hood should be located and designed so as to serve as a demonstration or work area. It may be placed in the advanced project area separated from the classroom-laboratory by a vision panel. In some schools, small down-draft hoods are used on the student laboratory tables preventing crowding around a single hood.
d. A still to provide "pure water" having a capacity of about one gallon per hour, is commonly mounted on a wall above a laboratory table; a shelf to support a 5-gallon receiving bottle should be provided. An ion exchange apparatus is used by some schools instead of a still.
e. Eye hooks are desirable in the ceiling above the demonstration table and in the perimeter ceiling above the lab area.
f. A work surface and storage for tri-beam balances, hand balances, and centrifuges should be provided. A work surface 10 feet long and 36 inches

high may be provided in the laboratory area, with the top of the same material as the demonstration desk and with overhang, no holes in the top, and free from obstruction. A locked cupboard with the drawer space below is desirable.
g. Chemical storage should be designed with special ventilation. It should be fire resistive and separate from storage space for apparatus and other supplies; it should be lockable. (See Appendix D.)
h. Provide an overhead shower where dangerous chemicals are stored, and an eye wash facility in or adjacent to the preparation room. Provide fire blanket, carbon dioxide fire extinguisher, first aid kit. Any radioactive materials should be stored in containers certified as safe by a recognized approving authority. Safety cabinets are available providing most of the above features. Such a cabinet should be located so as to be convenient to both pupils and instructors when needed.

Physics

a. For physics alone, normal ventilation should be sufficient and an acid resisting floor would not be necessary. Where chemistry also is to be taught, the statements under chemistry apply.
b. Eye hooks should be provided in the ceiling above the demonstration table and laboratory areas.
c. Physics laboratory tables that are movable are desirable for flexibility. Fixed service facilities may be located along perimeter walls or at islands in laboratory area.
d. The table top of the demonstration unit and tops of laboratory tables should have a 3-inch overhang to permit clamping of devices to the top, making the table tops completely free of holes and obstructions. Table tops should not have metal edges.
e. The demonstration table may be equipped to install a TV camera which can view small instrument experiments and transmit the image to TV receivers mounted above the demonstration table.
f. A conduit to the roof is desirable so that wire connections can be maintained to this outside area.
g. As a desirable accessory space, a science shop where electrical and mechanical repairs can be made for the entire science department may be provided.

Other desirable components of the science facilities are described below.

Multipurpose Science Room

Storage. Where only one science laboratory is served, at least 125 square feet of storage should be provided. A storage room serving two science rooms should have a minimum of 150 square feet, with 180 or more desirable. A central storage and preparation room serving three or more laboratories may require 400-500 square feet. Banks of graduated drawers and suitable open shelving are typical of general storage. One section of a storage room used for chemical storage should be provided with overhead shower and 12-inch shelving for dangerous chemicals. It should have safety ledges in front and be painted with acid resisting paint.

Office-conference-preparation Area. A combination office, conference, and preparation area is suggested, located adjacent to the storage area and with easy access to it. The office-conference area will provide space for desks and files for two instructors, clothing storage, conference space, and reference shelving. The preparation area should be provided with laboratory

Physical Plant and Facilities

counter, a sink with hot and cold water, soap and towel dispensers, gas and electric outlets, and additional provisions suited to the types of laboratories served. A safety cabinet with an eye wash fountain should be provided if a chemistry lab is served. Locate the safety cabinet in the science room adjacent to the preparation area, conveniently available to both instructor and pupils.

Advanced Project Area. Sufficient space should be provided to enable 6-8 pupils to work on special projects and research. A light-proof folding door may divide the space into two project areas, one for each of two adjacent science rooms, each with counter, water, gas, electric outlets and sink, work table, and research library.

Plant Room. (See Biology.)

Animal Room. (See Biology.)

Dark Room. A preferred location for a dark room is in the materials production center or audiovisual unit. In case of remodeling an older building in which a dark room is not provided elsewhere, it may be located adjacent to the general science or physics room.

Furniture and Equipment

General. The types of furniture and equipment for the classroom, laboratory, and accessory areas should be determined before working drawings are developed.

Chalkboard and Tackboard. Not less than 20 lineal feet of chalkboard and 16 lineal feet of tackboard, at least 42 inches high should be provided. The bottom of the chalkboard rail should be at least as high as the demonstration desk and counters (not less than 36 inches). Some science specialists recommend 20 lineal feet of chalkboard 48 inches high behind the demonstration desk, and above the chalkboard a map-tack rail for hanging charts. Still another map-tack rail should be provided at the wall-ceiling juncture for hanging charts 36 inches high so that they can be seen by the pupils. About 12 lineal feet of tackboard should be provided near the room entrance, and about 8 lineal feet of pegboard for mounting displays.

Display. In addition to tackboard and pegboard, provide display cases in the room and hall.

Other Details

Plumbing. Provide:
a. Water and gas connections for each laboratory area, each storage and preparation room, and each advanced projects room; all concealed waste lines acid resisting.
b. Master disconnect for all electric and gas services to demonstration desk, laboratory, and auxiliary rooms.
c. Vacuum breakers on all laboratory table and counter water supply pipes equipped with serrated tips from which the water outlets can be extended below the spill line by means of a hose. Note: About one-half of the gooseneck water supply pipes may be equipped with plain (nonserrated) tips which do not require vacuum breakers.

Ventilation. The ventilation system should provide adequate air change for science room operations. For general science and physics, 30 cfm per pupil with $7\frac{1}{2}$ cfm outside air. For biology, chemistry, and multipurpose science rooms, 8-10 air changes per hour, 100 percent outside air, 110 percent exhaust with manual switch control. This system supplements the normal ventilation system providing 30 cfm per pupil and $7\frac{1}{2}$ cfm outside air. Chemistry and biology storage spaces require 24-hour special exhaust ventilation. At least one fume hood should be provided in science rooms in which chemistry is taught.

Electrical.
  a. Outlets: Provide properly located electrical outlets for type, location, and arrangement of laboratory and audiovisual equipment; waterproof outlets for plant and animal rooms.
  b. Lighting: Provide a minimum of 50 foot-candles at work surfaces. Control artificial light in science rooms by dimmer switch at or near the demonstration desk.
  c. Daylight control: Daylight control devices permitting use of audiovisual equipment is necessary.

Sound Control. Provide acoustical planning of all science spaces.

Floors. Provide floors of a nonslippery type and of a material not affected by water and chemicals.

Fire Fighting Equipment

Fire extinguishers must be installed in all science laboratories. Extinguishers may be of the following types:

Soda acid
Water (stored carbon dioxide pressure cartridge)
Antifreeze solution (pressure)
Foam
Loaded stream

Provide not less than 15 pound CO_2 fire extinguisher and 15 pound dry powder extinguisher (ABC).

Provide at least one wool fire blanket in each laboratory installed in a metal cabinet identified with the words "fire blanket."

All extinguishers should be located so that a person will not have to travel more than 50 feet to reach the nearest fire extinguisher.

The fire extinguisher shall be properly inspected and maintained. An extinguisher should be recharged immediately after being used. All extinguishers, especially the soda acid and foam type, should be emptied and recharged once a year to prevent deterioration. Carbon dioxide fire extinguishers, including the cartridge of the dry powder type, loaded stream, and water and antifreeze solution stored pressure cartridge types, should be weighed at least once each year to detect loss by leakage. Any which show a loss of 10 percent or more of the rated capacity stamped on it, should be recharged.

Gas Service Lines

a. An outside gas shut-off valve, accessible, and plainly marked "MAIN GAS VALVE" should be installed on the building gas service main, preferably more than 5 feet from the building.
b. Main gas supply lines should not be placed under the building or in trenches or nonventilated spaces (such as suspended ceilings), unless properly cased and vented to the outside.
c. When gas outlets are provided in laboratories, a master control valve, quick closing, and *easily accessible* should be provided in each area where there are two or more outlets.

Miscellaneous

a. Curtains, draperies, decorative hangings, and similar materials should be noncombustible or flameproofed.

Physical Plant and Facilities

b. All flammable supplies and wastes should be stored in fireproof containers or in cabinets in locked storage rooms.
c. Heating units, such as gas plates and gas burners, should be insulated from table topes or other combustible materials.

   DUE TO POSSIBLE EQUIPMENT ABUSE, IT IS NOT RECOMMENDED THAT CLASSROOMS-LABORATORIES DESIGNED SPECIFICALLY FOR SCIENCE BE USED FOR INSTRUCTION IN OTHER CURRICULUM SUBJECT AREAS.

10

Science Safety for Handicapped Students

The passage of PL 94-142 and Section 504 of the Vocational Rehabilitation Act has greatly increased the interest of schools in the education of students who are handicapped. These laws mandate programming for handicapped students in the least restrictive environment appropriate and may result in the inclusion of many handicapped students in the regular curriculum and classroom. Many articles and textbooks have been written describing handicapped students and how curriculum can be adapted for them. An excellent source for describing various adaptations of science curriculum and equipment for the handicapped, as well as an extensive bibliography, is found in the proceedings of *A Working Conference on Science Education for Handicapped Students*, edited by Helenmarie Hofman.[1]

Articles and manuals have also been written describing the safety rules and procedures that need to be utilized to ensure a safe environment for students in the science classroom and laboratory. The problem of safety and the handicapped student has not been addressed in most of the materials reviewed. This oversight may be due in part to the fact that most handicapped children were excluded from most secondary (7-12) science courses, particularly the laboratory sciences. To quote Martha Ross Redden, a participant at the conference, "The response that came from the survey was indeed bleak, for there was not one state in the country that could define for us the science program for the physically handicapped youth."[2] This is not to say, however, that individual teachers have not accomplished much in educating handicapped students that have been in some of their classes. The inclusion of handicapped students in laboratory science courses increases the concern for their safety since they may not be able to see or hear or identify if they do see and hear potentially dangerous situations developing, and if the situation does become critical they may not be able to react rapidly enough to protect themselves.

Although it is not the intent of this chapter to discuss in detail how to teach handicapped students, some attention to this topic is necessary since most secondary teachers have had little experience with handicapped students. The following excerpt from Bybee[3] gives some guidelines teachers may wish to use.

---

    1. Helenmarie Hofman, ed., *A Working Conference on Science Education for Handicapped Students*, National Science Teachers Assn., April 1978, p. 4.
    2. Ibid., p. 4.
    3. Rodger Bybee, "Helping the Special Student Fit In," *Science Teacher* (October 1979): 23-24.

Understanding Special Students--General Guidelines

Certainly there are unique problems in integrating any special student into the classroom. Nevertheless, there are simple straightforward approaches that have proven helpful with most students:

--Obtain and read all the background information available on the student;
--Spend time educating yourself on the physical and/or psychological nature of the handicap, and how it affects the student's potential for learning;
--Determine whether or not special help can be made available to you through the resources of a "special education" expert;
--Determine any special equipment needed by the student;
--Talk with the student about limitations due to his or her handicap and about particular needs in the science class;
--Establish a team of fellow teachers (including resource teachers and aides) to share information and ideas about the special students. A team approach is helpful in overcoming initial fears and the sense of aloneness in dealing with the problem. You may need to take responsibility for contacting appropriate school personnel and establishing the team;
--Other students are often willing to help special students. Use them;
--Be aware of barriers--both physical and psychological--to the fullest possible functioning of the special student;
--Consider how to modify or adapt curriculum materials and teaching strategies for the special student without sacrificing content;
--Do not underestimate the capabilities of the special student. Teachers' perceptions of a student's abilities have a way of becoming self-fulfilling prophesies. If these perceptions are negative, they may detrimentally affect the student and your ability to create new options for him or her.
--Use the same standards of grading and discipline for the special student as you do for the rest of the class;
--Develop a trusting relationship with the special student;
--Educate the other students about handicaps in general, as well as specific handicaps of students in their class.

Hearing Impaired

--The hearing impaired depend heavily on visual perception. Therefore, seat the student for optimal viewing;
--Determine whether an interpreter will be needed and the nature of the child's speech/language problems;
--Learn the child's most effective way of communicating;
--Find the student a "listening helper."

Visually Impaired

--Visually impaired students learn through sensory channels other than vision, primarily hearing. Therefore, seat students for optimal listening.
--Determine from the student what constitutes the best lighting;
--Change the room arrangement whenever necessary, but always make a special effort formally and informally to reorient the student;
--Allow the student to manipulate tangible materials, models, and when possible "real" objects. Do not unduly "protect" students from materials;
--Speak aloud what you have written on the board and charts;

- --Use the student's name; otherwise, the student may not know when he or she is being addressed;
- --Since smiles and facial gestures might not be seen, touching is the most effective means of reinforcing the student's work;
- --Be aware of student eye fatigue. This can be overcome by varying activities, using good lighting, and providing close visual work.

Physically and Health Impaired
- --Eliminate architectural barriers;
- --Become familiar with the basic mechanics and maintenance of braces, prostheses, and wheelchairs;
- --Understand the effects of medication on students and know the dosage;
- --Obtain special devices such as pencil holders or reading aids for students who need them;
- --Learn about the symptoms of special health problems, and appropriate responses.

Speech and Language Impaired
- --Help the student become aware of his or her problem; students must be able to hear their own errors;
- --Incorporate and draw attention to newly learned sounds in familiar words;
- --Know what to listen for, and match appropriate remediation exercises with the student's problem;
- --Be sure your speech is articulate; students often develop speech and language patterns through modeling.

Learning Disabilities and Mental Retardation
- --Listen closely so you can understand the student's perception and understanding of concepts and procedures;
- --Use an individualized approach based on the student's learning style, level of understanding, and readiness;
- --Use multisensory approaches to learning: visual, auditory, kinesthetic, and tactile;
- --Find and use the student's most refined sensory mode to aid in development of mental capacities;
- --Teach to the student's strengths, and work on diminishing his or her deficiencies;
- --Reduce or control interruptions since many special students have short attention spans;
- --Stay within the student's limits of frustration. Rely on your judgment, not the level of curriculum materials;
- --Start conceptual development at a sensory-motor or concrete level, and work toward more abstract levels;
- --Work on speech and language development;
- --Help special students to develop self-esteem; a good, firmly grounded self-concept is essential to their continued development.

Emotionally Disturbed and Disruptive Students
- --Make rules reasonable and clear;
- --Provide realistic, appropriate consequences if rules are broken;
- --Never use physical punishment for rule violation;
- --Disruptive behavior ranges from low levels at which a student may merely be looking for attention or recognition through a spectrum that ends in rage, tantrums, or complete withdrawal. Try always to be alert to

behaviors that, though minimally disruptive, could become more serious problems;

--Resolve conflicts by talking and reasoning with the student. Once a course toward aggressive or uncontrolled behavior is started, it is hard to stop;

--If behavior problems escalate, try to talk about the process while providing ways out of the problem. For example, "We are both getting angry, can't we settle this calmly," or "I see you are upset, let's try to solve the problem."

What type of facility for the laboratory will accommodate the handicapped? One source indicates, "It can be seen that a laboratory serving students with physical disabilities differs from a conventional one in respect to allowing more space per pupil, and providing accessible work stations and storage areas."[4] One example of a facility for the laboratory which will accommodate the physically impaired is found in *Barrier-Free School Facilities for Handicapped Students*. To quote, "At least one student laboratory station should be provided to accommodate the physically impaired. This would necessitate that a 30-inch clearance exists between the bottom of the work surface or apron and the floor. This space must also be 18-inches deep and a minimum of 30-inches wide. Hot water pipes in this space should be insulated. Faucets and utility outlets at these stations should be side-mounted rather than rear-mounted and be equipped with wristblade or lever handles. The aisles in this area should be a minimum of 36 inches wide and clear of obstructions."[5]

Yuker, Revenson, and Fracchia suggest the following design considerations for a science laboratory: work areas mounted at wheelchair height along the perimeter of the room; a student work area of 4 feet with 8 feet between sets of gas, water, and electrical outlets; storage alongside the work area and beneath the work surface and a sink installed with the lowest edge at a height of 27 inches with recessed plumbing. Faucets with gooseneck spigots and batwing taps are suggested since they require less strength and dexterity.[6] A portable laboratory bench has been designed by the Department of Chemistry, University of California, Davis. A commercially produced portable science station for the physically handicapped is available from Conco Industries, Inc.

The previously described facilities, although making it easier for the handicapped student to function, do present some concerns for the safety of the student. The students using these desks will generally be sitting and thus become exposed to several potentially dangerous situations. First, the student, due to sitting, has the apparatus and equipment at face level. Second, any spills have the potential to end up in the student's lap. To protect the student's face, neck, chest, and lap it is recommended that the student be provided with an approved full face and neck shield and appropriate body cover depending upon the handicap.

Consideration must also be given to students who may have prosthetic limbs, artificial hands, hooks, or other artificial body parts. In addition, students who are subject to seizures or who, from their handicap, tire easily

---

4. Harold E. Yukers; Joyce Revenson; and John F. Fracchia, *The Modification of Education Equipment and Curriculum for Maximum Utilization by Physically Disabled Persons*, ED 031022, 1968, Human Resources Center, Albertson, N.Y., p. 16.

5. ———, *Barrier-Free School Facilities for Handicapped Students*, Educational Research Service, Inc., 1977, p. 75.

6. Ibid., pp. 16, 32, 33.

must also be provided for. Some or all of these students may need help in getting in and out of their protective clothing and equipment. A few may need someone to sequence the activities so that they can see how to use and practice the procedures necessary to complete the assigned activity. A stool should be provided for students who become easily fatigued. It is recommended that a bar be placed across the front of the laboratory station so that students who may have a seizure or become suddenly fatigued will not fall directly into the apparatus on the laboratory table.

In her work, Hofman indicates that the handicapped student must strive to gain independence in the laboratory situation. This will be facilitated if separate accessible shelves are provided for their use, labels are provided in Braille, if necessary, and electric hot plates replace Bunsen burners if possible.[7]

Another technique suggested by Hofman is that the handicapped student be paired with a "normal" student or another handicapped student whose strengths can compensate for the other's handicap.[8] The Hofman work describes a procedure for blind students to use in safely lighting a Bunsen burner. In addition, Hofman describes how blind students can use hypodermic syringes to measure and pour corrosive chemicals.[9]

Obviously, the handicapped student may not be able to perform every experiment that other students can do. However, the following quote sums up general feelings on the subject. "Occasionally a modification will need to be made in assignments, such as using a microscope. Such modifications should be the exception rather than the rule, however; and when really necessary, the modification should provide a logical equivalent, rather than a substitute which is easier and/or irrelevant."[10] The science teacher's commonsense has to be the final judge of what is or is not allowed while still keeping the intent of PL 94-142 and Section 504.

This discussion cannot include all the many procedures and methods that can be used to modify experiences for the handicapped. The reference list includes more sources of information. (See also Appendix A, p. 113.)

---

7. Hofman, p. 168.
8. Ibid.
9. Ibid., p. 97.
10. Doris Willoughby, ed., *Your School Includes a Blind Student*, National Federation of the Blind Teachers Division, Chatsworth, Calif., p. 22.

REFERENCES

American Association for the Advancement of Science. *Science Education News*. Washington, D.C.: AAAS Fall 1978/Winter 1979. (Contains large bibliography.)

American Foundation for the Blind, Inc., 15 West 16th Street, New York, N.Y. 10011.

American Printing House for the Blind, Inc., 1839 Frankfort Avenue, Louisville, Ky. 40206.

Hofman, Helenmarie H., *A Working Conference on Science Education for Handicapped Students*, National Science Teachers Association, April 1978. (Contains large bibliography.)

Hofman, Helenmarie H., and Ricker, Kenneth S., *Science Education and the Physically Handicapped*, National Science Teachers Association, Washington, D.C., 1979.

National Technical Institute for the Deaf, One Lomb Memorial Drive, Rochester, N.Y. 14623.

Physics Teacher, Dec. 1973, Sept. 1977.
Science and Children, March 1976, Nov.-Dec. 1976, Jan. 1977.
Science for the Blind, Inc., Box 120, Bala-Cynwyd, Pa. 19004.
Science Teacher, April 1973, Dec. 1974, Dec. 1975, Dec. 1978.
Willoughby, Doris, ed., *Your School Includes a Blind Student*, National Federation of the Blind Teachers Division, Chatsworth, Calif.
Yukers, Harold E.; Revenson, Joyce; and Fracchia, John F., *The Modification of Education Equipment and Curriculum for Maximum Utilization by Physically Disabled Persons*, ED 031022, 1968, Human Resources Center, Albertson, N.Y.

Accident/Incident Reporting Systems

ACCIDENT REPORTS

Accident data may be put to a great many uses. The information available through an effective school accident reporting system can be used by teachers, school nurses, custodians, building principals, department heads, school superintendents, school boards, school district attorneys, and many others.

Specifically, the systematic accumulation of school accident and injury data will provide information upon which to base:

1. Curriculums designed to educate the child for safe living.
2. A realistic evaluation of safety program efforts on a regular basis.
3. Changes in buildings, facilities, equipment or procedures to improve the environment of the school system.
4. Organizational and administrative improvements to strengthen the managerial aspects of the safety program.
5. A strong public relations program, thus lessening public demands for crash programs of little value if an unusual incident occurs.
6. An assessment of the costs of accidents and injuries and their relationships to the operating expenses of the school system.

Individual accident reports can be useful for positive action at each of several levels of the school system. Therefore, reports should be reviewed carefully at each level through which they are processed. Generally, an individual accident report:

1. May show conditions or deficiencies which can be immediately corrected.
2. Can be used as a teaching tool.
3. Can strengthen staff interest in accident prevention activities through review of the accident by appropriate persons.

In addition, individual accident reports can provide important information in cases involving teacher or school district tort liability.

Many different accident reporting systems can be developed. Any system is good if it gives valid results in the simplest form and with the least amount of effort. The answers to three questions will provide assistance in the development of an effective accident reporting system.

Adapted from the National Safety Council, *Student Accident Reporting*, 1966.

1. *Which accidents should be reported?*
    In most schools, a reportable accident is any school jurisdictional accident which results in any injury to a pupil and/or property damage (school jurisdiction includes school building, grounds, to and from school, and school-sponsored activities away from school property; the concept of property damage includes damage to the school's own equipment, material, or structures).
    Every school system should require a report of some kind on accidents defined as reportable. It is essential that the system knows of all accidents occurring under the school's jurisdiction. This sound practice offers some protection for the school in those areas where liability may be a factor. Most importantly, educational and preventive action can only be accomplished if it is known where and how all accidents and injuries are occurring.
2. *What information should an accident report form provide and what form should it take?*
    No one report form will meet the needs of every school system. There is, however, a required body of information basic to the effective analysis and utilization of accident and injury data. Required information includes:

    a. Name of person involved
    b. Home address
    c. School
    d. Sex
    e. Age
    f. Grade/Special Program
    g. Date and time of the accident, day of the week
    h. Nature of injury
    i. Degree of injury
    j. Cause of injury
    k. Location of accident
    l. Activity of person
    m. Supervision (who was in charge?)
    n. Agency involved (apparatus, equipment, etc.)
    o. Unsafe act
    p. Unsafe mechanical--physical condition
    q. Unsafe personal factor (bodily defects, lack of skill, etc.)
    r. Description of the accident
    s. Date of report
    t. Report prepared by (signature)
    u. Principal's signature

    There is no model format. Keep it simple, easy to read, easy to complete, and provide enough writing room for a complete description of the incident.
    Here are some considerations and time-saving ideas.

a. The form can either be mimeographed or printed, depending on the number needed.
b. Most schools use an 8½ x 11 form; others use an 8½ x 5½ card. Some use both sides of the sheet or card. The form must be large enough to handle easily.
c. If several copies are needed, the use of printed carbon impregnated forms is practical.
d. Paper should not be so thin as to tear when an erasure is made.
e. Some school systems use check-off blocks for certain items. This may or may not be a time saver; it may be easier to simply write in the appropriate item.

3. *Who will use the form?*

The person in charge of the activity at the time of the accident is the logical person to initiate the accident report. Placing the responsibility on a third party who may have no firsthand knowledge of the incident, would tend to make the report less usable. However, reports from witnesses, particularly those individuals directly involved, may supply additional helpful information.

After initial completion of the report, it should be reviewed by the principal, and as appropriate, by other school personnel, prior to forwarding to the superintendent's office for action and filing. Examples of school personnel who may be able to use or contribute to the report are the school nurse, the custodian--particularly if there is an unsafe mechanical or physical condition involved--a department head, the school safety coordinator, or the safety committee. It may be helpful to provide copies for some or all of those listed.

Monthly and annual summaries of accident report forms can provide the information needed to direct the thrust of a school district-wide program to aid in insuring that students may learn in facilities which are as hazard-free as we can make them.

INCIDENT REPORTS

Individual science teachers or department heads may wish to keep records or file reports on near-accidents, facility and equipment defects, and/or unusual events which may have future impact on the participants of the science program.

The reports or records might serve a useful purpose in providing a safe learning environment or in cases involving teacher liability. (See also Appendix A, p. 113.)

APPENDICES

A
Checklists

LEGAL LIABILITY CHECKLIST

1. Teachers understand the following legal concepts as they relate to science teaching:

    Negligence   _____
    Due Care   _____
    Reasonable and Prudent Judgment   _____
    Tort Liability   _____
        Save Harmless Provision   _____
        Insurance   _____
    Good Samaritan Law   _____
    Sovereign Immunity Law   _____

2. Teachers are aware of state and local provisions related to:

    Adequate Supervision   _____
    Appropriate Instruction   _____
    Equipment Maintenance   _____
    Student Protective Equipment   _____
    Fire Codes   _____
    Emergency Procedures   _____
        Personal Injury   _____
        Evacuation Procedures   _____

EYE PROTECTION CHECKLIST

1. Teachers are aware of the appropriate ANSI specifications and usage of the following face protective equipment:

    Spectacles   _____
    Cover goggles   _____
    Face shields   _____
    Laser spectacles   _____

2. Teachers know how to identify eye protective equipment which meets ANSI standards.   _____

3. Teachers know that contact lenses do not provide adequate eye protection and shall not be worn in a hazardous environment without appropriate cover goggles. _____
4. Teachers know that laboratory visitors must wear appropriate eye protection when hazardous situations exist in the lab. _____
5. Teachers understand the Iowa School Eye Protection Law, and enforce it in their laboratories. _____
6. Teachers incorporate appropriate disinfection procedures when cleaning eye hardware. _____
7. Eye and face protective equipment is stored in clean, dust-proof cabinets. _____

EYE CARE CHECKLIST

1. Teachers encourage all students enrolling in science courses to have professional eye examination or visual screening by school nurse or eye physician. _____
2. Ascertain names of all students wearing contact lenses. _____
3. Check all students' corrective eye wear to determine if they meet Z 87.1 specifications or are dress (street) wear. _____
4. Supply or have each student purchase approved 6-vent cover goggles. _____
5. Require constant wear of approved 6-vent cover goggles during all experiments. _____
6. Emergency care supplies:

   Approved eyewash fountain in laboratory area _____
   Q-Tips, ample supply _____
   Sterile eye patches, ample supply _____
   Surgical tape, ample supply _____
   Eye cups, ample supply _____
   Posted WATS phone number of the University of Iowa Poison Control Center _____
   Posted phone number of ophthalmologist or hospital _____
   Posted phone number of emergency transportation _____

7. Teachers should know emergency first aid procedures for the following:

   Lids (contusions, lacerations) _____
   Eyes (contusions, penetrations) _____
   Eyes (chemical burns) _____

SAFETY IN BIOLOGY SETTINGS CHECKLIST

1. Storage

   All volatile liquids are stored in designated area away from classroom or lab. _____
   Drugs are stored in locked cabinets. _____

Preserved specimens are properly stored and accessible. _____
Special areas are provided for holding live animals. _____
Bulk storage areas are regularly inventoried. _____
Bulk storage areas are kept locked. _____

2. Cleanliness

   Animal cages are cleaned daily. _____
   Glassware utilized for microorganisms is sterilized
   prior to washing. _____

3. Disposal

   Special covered containers are provided for disposal of
   discarded biological materials. _____
   Waste biological materials are burned or buried. _____
   Waste chemicals, used for preservations, are disposed of
   by washing down drain with copious amounts of water. _____
   Disposable petri dishes are incinerated. _____
   Waste chemicals are properly disposed of according to the
   chapter on "Chemical Aspects of Safety" or the "Laboratory Waste Disposal Manual," MCA. _____

4. Microorganisms

   Known pathogens are *not* utilized in laboratory studies. _____
   All cultures are treated as though they contain pathogens. _____
   Media used for culturing microorganisms is sterilized. _____
   Inoculating loops are carefully flamed prior to transferring cultures. _____
   Rubber gloves are always worn while working with cultures. _____
   Laboratory areas used for microbiology are cleaned
   regularly. _____

5. Poisonous plants

   Students are instructed in the identification of poisonous plants indigenous to their area. _____

6. Animals in the laboratory

   Poisonous animals are *not* kept in the laboratory. _____
   Diseased animals are removed from the laboratory. _____
   Leather gloves are *always* worn when handling animals. _____
   Experiments involving animals are regularly monitored
   to assure humane treatment. _____

7. Killing jars for insect studies

   Carbon tetrachloride and cyanide compounds are *not* used
   as killing agents. _____
   Killing jars are clearly labeled as such and include
   the killing agent. _____

8. Preserved plant and animal materials

> All preservative chemicals are clearly labeled. _____
> Prior to dissection, specimens are thoroughly rinsed to
> remove excess preserving chemicals. _____

SAFETY IN CHEMISTRY SETTINGS CHECKLIST

1. Storage

> All chemicals are secure. _____
> Chemical groups are properly isolated from those with
> which they might react. _____
> Flammable liquids are stored in OSHA approved cabinets. _____
> There are no ignition sources near flammable liquid
> storage areas. _____
> Appropriate safety items (aprons, gloves, shields) are
> kept near areas where reactive chemicals are stored. _____
> Extremely hazardous and carcinogenic chemicals have been
> eliminated from the science program. _____
> Compressed gas bottles are secured and protected with a
> screw cap. The dates of receipt and inspection have
> been recorded. _____
> Fume hood is not used for storage of chemical reagents. _____
> Food is not stored in laboratory refrigerator. _____
> Appropriate flammable reagents are stored in explosion-
> proof or explosion-safe refrigerator. _____
> All chemical hardware is properly stored to prevent
> accidents during emergencies. _____
> There are no excessive quantities of hazardous chemical
> reagents being stored. _____
> Plastic, metal or other safety containers are being used,
> where appropriate, in lieu of glass. _____
> The flammable chemical storage area is in compliance with
> NFPA standards. _____
> Acid, base, and flammable liquid clean-up supplies are
> conveniently located for emergency usage. _____

2. Handling

> Waste jars are accessible to students for solids that are
> not to be disposed of down a sink. _____
> Adequate fume hoods are provided for chemical experiments
> conducted in the laboratory. _____
> A suitable system for safe dispensing of chemicals to stu-
> dents is in operation. _____
> Beaker tongs, suction bulbs for pipetting, and other ap-
> propriate lab hardware are available. _____
> Appropriate reference manuals are readily available in
> laboratories dealing with safe disposal of chemicals. _____
> Unlabeled/unidentified chemicals have been removed from
> the laboratory. _____

Checklists 111

3. Equipment

    Fire extinguishers are strategically placed for emergency usage in laboratory, classroom, and storage areas. _____

    Eye wash and deluge showers are checked regularly to insure proper operation. _____

    Fire blankets are strategically placed in the laboratory and storage areas. _____

4. Inventory

    An inventory system is regularly used that includes quantity, purchase date, receiving date, and date container first opened. (Person in charge of inventory is _____). _____

SAFETY IN PHYSICS SETTINGS CHECKLIST

1. Know procedure to follow when an accident has occurred. _____
2. Have a list of persons to contact for assistance when an accident has occurred. _____
3. Fire extinguishers are properly located and in working order. _____
4. Fire blankets are located in the laboratory area. _____
5. Flammable liquid stored in OSHA approved cabinets. _____
6. Laboratory gas valves are in good working condition. _____
7. Know the location of electrical circuit breakers to shut off electricity to the laboratory. _____
8. Know the location of the main shut-off valves for gas and water in the laboratory. _____
9. First aid kit is in proper location. _____
10. Have proper personal protective devices for: Eye/Face _____
    Hands _____
    Hair _____
11. All 110 volt AC outlets are equipped with ground fault interrupters if possible. _____
12. High voltage equipment has interlocks. _____
13. Areas where experiments using 110 volt AC with exposed connections are performed are a safe distance from gas and water fixtures. _____
14. Isolation transformers are available for working with 110 volt AC experiments. _____
15. Check all electrical (110V AC) devices for proper grounding and faulty insulation. _____
16. Lasers are properly encased to prevent shock and have key operated switches. _____
17. Radioactive materials have proper storage. _____
18. Mercury is stored in tightly closed containers. _____
19. Motorized equipment has proper guards. _____
20. Belts on motorized equipment are in good condition. _____
21. Hot plates are available for heating flammable liquids. _____
22. Lead storage batteries stored with proper ventilation. _____

FIELD ACTIVITIES CHECKLIST

1. Teacher has visited proposed site and has knowledge of the potential hazards and areas to be avoided _____

       Inspect site for: hazardous trails; _____
                        poisonous plants; _____
                        dangerous animals. _____
       Instruct students and supervisors concerning hazards and safe procedures. _____
       Provide appropriate safety equipment. _____
       Be prepared to contact emergency personnel. _____
       Instruct students concerning appropriate clothing. _____
       Instruct students concerning protection of nature. _____

2. Proper consent and permission forms have been obtained (parents, school authorities, property owners). _____
3. Provisions have been made for adequate supervision. _____
4. Appropriate *school*, or school endorsed, transportation has been arranged. _____

STUDENT RESEARCH CHECKLIST

1. Teacher has evaluated the research to ensure appropriate safety measures have been planned for the topic. _____
2. Student has sought appropriate expertise in setting up the experimental design. _____
3. Student is aware of appropriate codes governing the use of scientific materials (humane treatment of animals is one example). _____
4. Appropriate safety procedures have been planned for the investigator and others. _____
5. Appropriate supervision has been planned. _____
6. The teacher and student have planned appropriate safety measures for any projects to be done in part or totally at home. _____

PHYSICAL PLANT AND FACILITIES CHECKLIST

1. Chemical storage areas are designed with special ventilation, to be fire resistive, separate from other supply areas, and lockable. _____
2. An overhead deluge shower is provided in the laboratory. _____
3. Chemical storage shelves are 12 inches deep, have safety ledges in front, and are painted with acid resistant paint. _____
4. Adequate numbers of fume hoods are provided in laboratory areas for experimental purposes. _____
5. Adequate numbers of eye washes are strategically placed for emergency usage in laboratories. _____
6. Master control valves for electricity and gas are readily accessible and located away from chemical storage areas. _____
7. Flammable and explosion proof storage cabinets are utilized where necessary. _____

8. Sufficient electrical service is available.
9. Adequate ventilation in science rooms (general ventilation as opposed to specific vent).
10. Adequate lighting in laboratory areas as well as lecture or classroom areas.
11. Is science safety equipment functional?
12. Are emergency evacuation routes posted?
13. Is the science program accessible to the handicapped person?
14. Are facilities modified to accommodate the handicapped person, i.e. lab stations, auxiliary facilities, etc.?

HANDICAPPED STUDENTS IN SCIENCE CHECKLIST

1. Teachers plan orientation sessions for parents and handicapped pupils.
2. Teachers understand PL 94-142, Section 504 as it relates to science teaching.

ACCIDENT/INCIDENT REPORTING CHECKLIST

1. Accidents/incidents occurring in the science area are reported in writing to designated school officials.
2. An effective, complete accident/incident reporting form is available for use in the school district.
3. Accident/incident reports are regularly reviewed in efforts to improve safety in the science setting.
4. Copies of accident/incident reports are kept in teacher files.

B
Eye Protectors

Eye Protectors 115

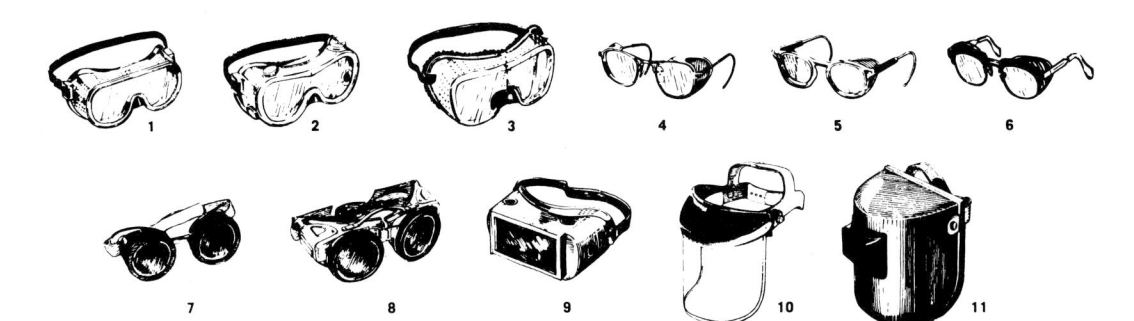

1. **GOGGLES,** Flexible Fitting, Regular Ventilation
2. **GOGGLES,** Flexible Fitting, Hooded Ventilation
3. **GOGGLES,** Cushioned Fitting, Rigid Body
*4. **SPECTACLES,** Metal Frame, with Sideshields
*5. **SPECTACLES,** Plastic Frame, with Sideshields
*6. **SPECTACLES,** Metal-Plastic Frame, with Sideshields
** 7. **WELDING GOGGLES,** Eyecup Type, Tinted Lenses (Illustrated)
7A. **CHIPPING GOGGLES,** Eyecup Type, Clear Safety Lenses (Not Illustrated)
** 8. **WELDING GOGGLES,** Coverspec Type Tinted Lenses (Illustrated)
8A. **CHIPPING GOGGLES,** Coverspec Type, Clear Safety Lenses (Not Illustrated)
** 9. **WELDING GOGGLES,** Coverspec Type, Tinted Plate Lens
10. **FACE SHIELD** (Available with Plastic or Mesh Window)
11. **WELDING HELMETS

*Non-sideshield spectacles are available for limited hazard use requiring only frontal protection.
**See appendix chart "Selection of Shade Numbers for Welding Filters."

APPLICATIONS

OPERATION	HAZARDS	RECOMMENDED PROTECTORS: Bold Type Numbers Signify Preferred Protection
ACETYLENE–BURNING ACETYLENE–CUTTING ACETYLENE–WELDING	SPARKS, HARMFUL RAYS, MOLTEN METAL, FLYING PARTICLES	7, 8, 9
CHEMICAL HANDLING	SPLASH, ACID BURNS, FUMES	**2**, 10 (For severe exposure add **10** over **2**)
CHIPPING	FLYING PARTICLES	**1, 3,** 4, 5, 6, **7A, 8A**
ELECTRIC (ARC) WELDING	SPARKS, INTENSE RAYS, MOLTEN METAL	**9, 11** (**11** in combination with **4, 5, 6,** in tinted lenses, advisable)
FURNACE OPERATIONS	GLARE, HEAT, MOLTEN METAL	**7, 8, 9** (For severe exposure add **10**)
GRINDING–LIGHT	FLYING PARTICLES	**1, 3,** 4, 5, 6, 10
GRINDING–HEAVY	FLYING PARTICLES	**1, 3, 7A, 8A** (For severe exposure add **10**)
LABORATORY	CHEMICAL SPLASH, GLASS BREAKAGE	**2** (10 when in combination with **4, 5, 6**)
MACHINING	FLYING PARTICLES	**1, 3,** 4, 5, 6, 10
MOLTEN METALS	HEAT, GLARE, SPARKS, SPLASH	**7, 8** (**10** in combination with **4, 5, 6,** in tinted lenses)
SPOT WELDING	FLYING PARTICLES, SPARKS	**1, 3,** 4, 5, 6, 10

Appendix Fig. B.1. Selection Chart for Eye and Face Protectors for Use in Industry, Schools, and Colleges. (Reprinted by permission from American National Standard Practice for Occupational and Educational Eye and Face Protection, ANSI Std. Z 87.1-1968, American National Standards Institute.)

C

Explosion-safe Refrigerators

Explosion-safe, flammable storage refrigerators (12 cu ft models) are available from Precision Scientific Co. Current price is approximately $945 (1980).

Conventional home refrigerators (certain models) can be modified to be made explosion safe. Only manual defrosting refrigerators should be used since the automatic defrosting creates an ignition source difficult to isolate. These manual defrost units are also cheaper and cost less to operate. For example, a 12 cu ft manual defrost refrigerator requires an average of 47 KW/month whereas the 12 cu ft automatic defrost requires 95 KW/month. At the current average rate for electricity the cost for manual defrost would be $3.52/month vs $7/month for automatic defrost.

A modified, conventional home refrigerator costs approximately $600 less than the retail cost of an explosion-safe flammable storage refrigerator. To make this manual defrost home refrigerator explosion-safe requires the removal of all sources of ignition from the interior storage compartment and relocating the control to the exterior of the box. All penetrations through the interior wall should also be sealed. The following Appendix Figures C.1 to C.5 illustrate the modification of a typical manual defrost home refrigerator.

Explosion-safe Refrigerators   117

Appendix Fig. C.1. (1) Cover plate where control was removed; (2) Sealed light bulb receptacle; (3) Temperature control probe.

Appendix Fig. C.2. (1) Temperature control; (2) Temperature control probe penetrating the wall; (3) Wiring from the control to the compressor.

Appendix Fig. C.3. Silicone sealing compound.

Appendix Fig. C.4. Door switch removed and the hole plugged.

Explosion-safe Refrigerators 121

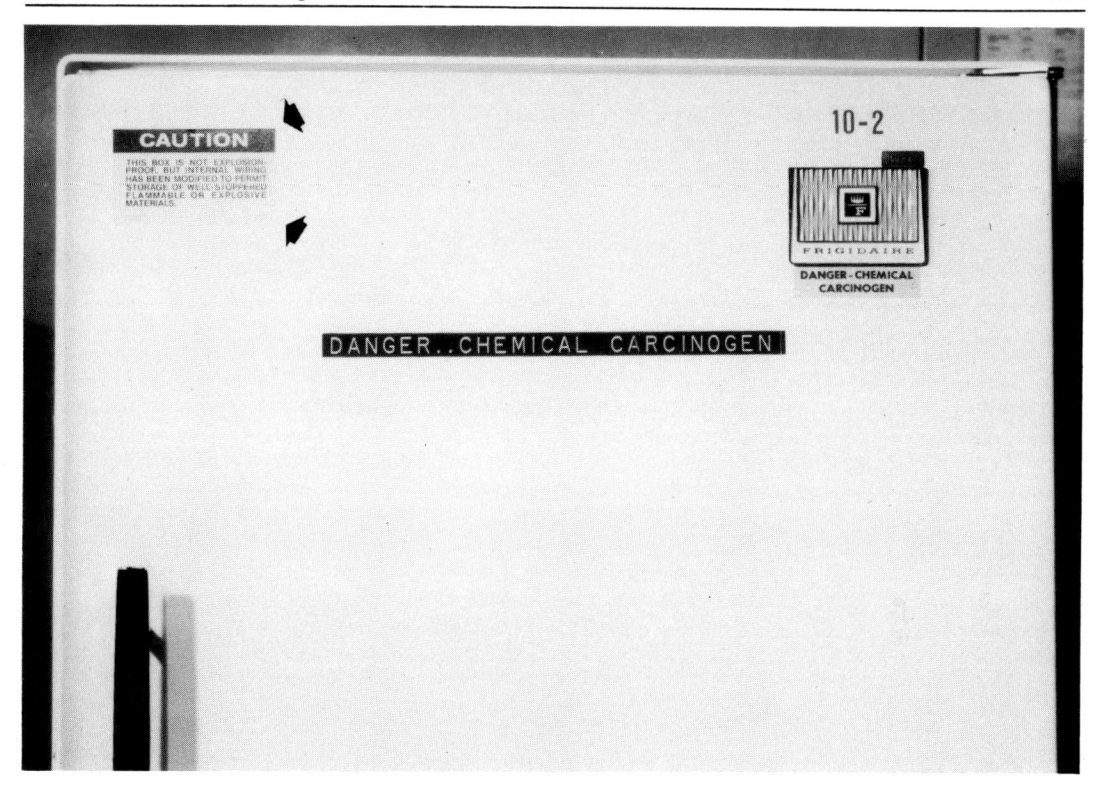

Appendix Fig. C.5. Properly labeled explosion-safe refrigerator.

D
Safety Equipment

Safety Equipment 123

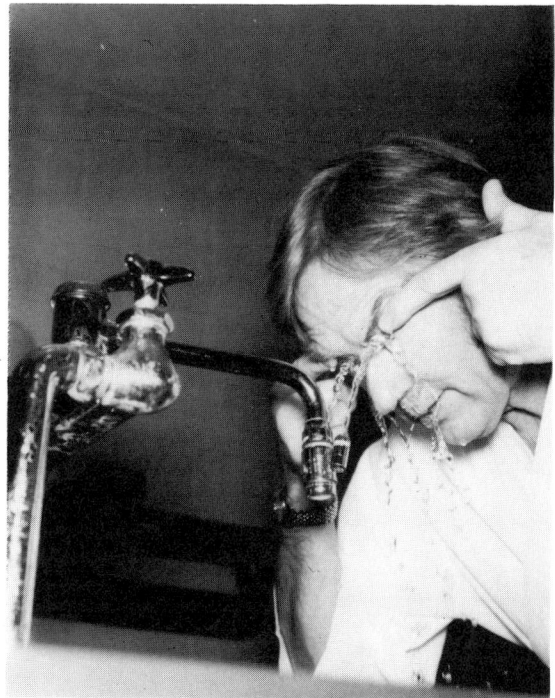

Appendix Fig. D.1. Emergency Eye Wash Fountain. (Obtainable from Lab Safety Supply Co., Janesville, Wisconsin.)

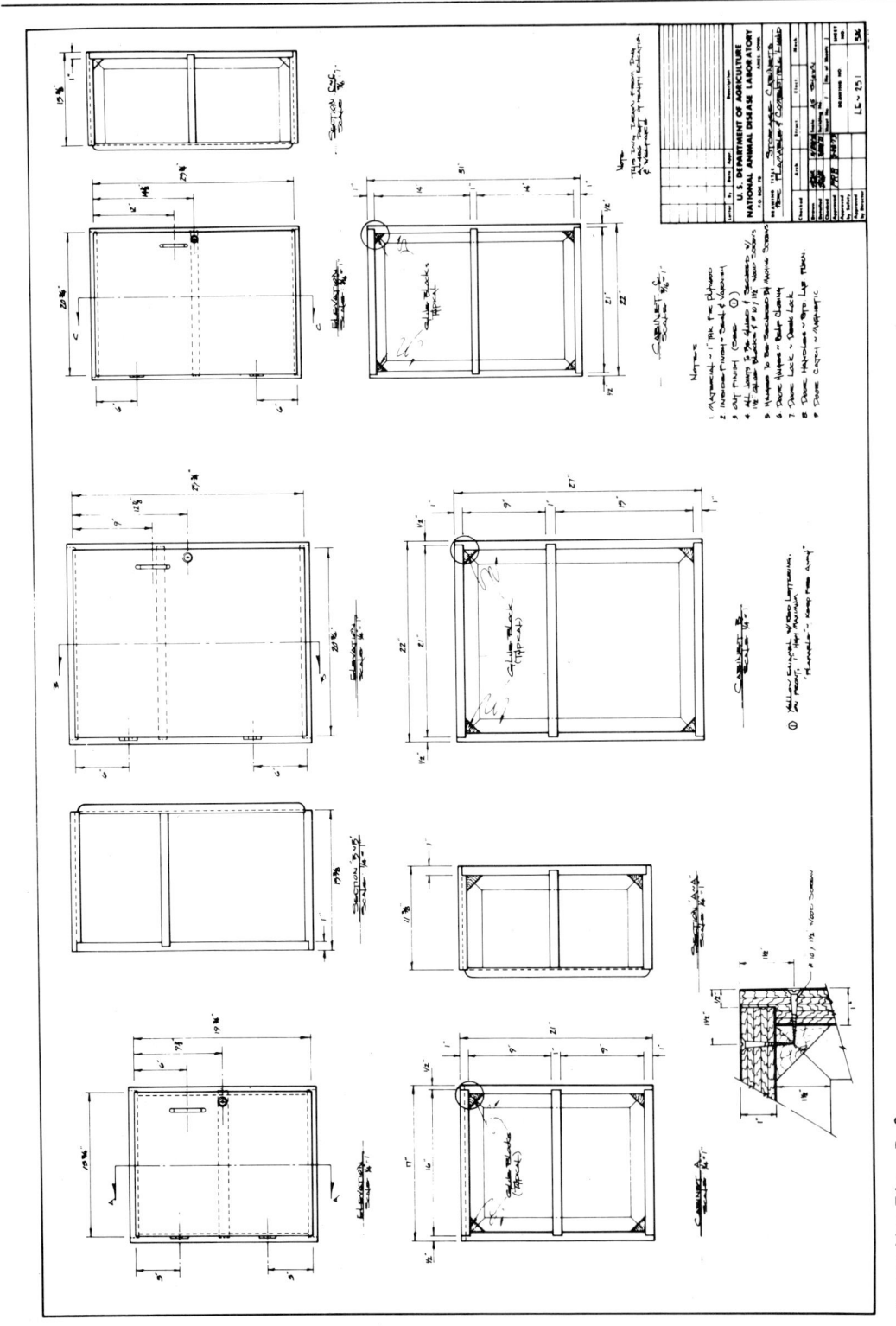

Appendix Fig. D.2.

Safe, Convenient, Portable Pipettor

J. R. SONGER

National Animal Disease Center, Agricultural Research Service, Ames, Iowa 50010

Received for publication 15 November 1977

A safe, convenient, portable pipetting device that will accommodate any size pipette is described. A vacuum bulb eliminates the need for external vacuum. Necessary components, fabrication procedures, and operating techniques are given.

Many laboratory workers still pipette by mouth. The hazards associated with this procedure have been clearly documented (2, 5). Mouth pipetting has become such a habit with most laboratory workers that change comes with great reluctance. Most commercially available pipetting devices (1) contribute to this reluctance because of their inconvenience. Some of these devices must be attached to and removed from the pipette with each pipette change. Other devices fit pipettes of only one size, and the device must be changed each time a pipette size is changed. With some devices fingers become cramped and dexterity is lost if the device is used for more than a few minutes. Pipetting devices that I previously described (3, 4) were designed to minimize these difficulties.

The pipettor described in this note was designed to incorporate the convenience of a finger pipettor (3) with a hand-held vacuum source. This unit will accommodate pipettes of all sizes. Only the vacuum level must be adjusted when pipette sizes are changed.

The fabrication procedure follows: pipettor components are shown in Fig. 1. Fabrication of the finger-mounted part of the pipettor, which

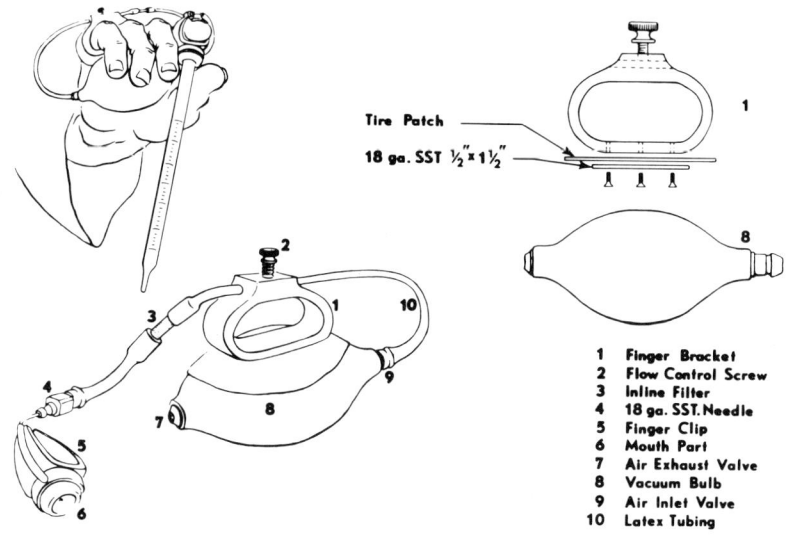

FIG. 1. *Safe, convenient, portable pipettor. A tire patch was bolted to the finger bracket and then cemented to the vacuum bulb. The latex tubing was attached to the air inlet valve and threaded through the hole in the finger bracket. The latex tubing was cut and the in-line filter was inserted in the tubing. The free end of the latex tubing was attached to the 18-gauge, stainless-steel needle of the finger pipettor assembly.*

1. Finger Bracket
2. Flow Control Screw
3. Inline Filter
4. 18 ga. SST. Needle
5. Finger Clip
6. Mouth Part
7. Air Exhaust Valve
8. Vacuum Bulb
9. Air Inlet Valve
10. Latex Tubing

Appendix Fig. D.2. Suggestions for the construction of an approved cabinet (wood) for storage of flammable liquids. A continuous hinge (piano hinge) can be mounted so that screws are not exposed when the door is closed. Cabinet can be painted yellow with red lettering: FLAMMABLE--KEEP FLAME AWAY.

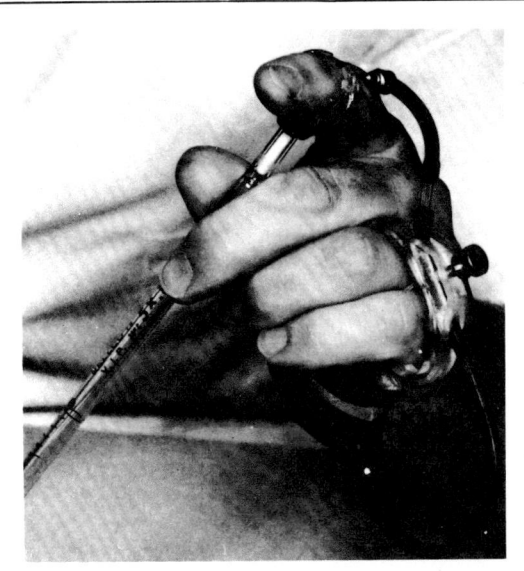

FIG. 2. *Operation of the safe, convenient, portable pipettor.*

consists of parts 4, 5, and 6 shown in Fig. 1, has been described by Songer et al. (3).

The finger bracket (Fig. 1, part 1) was cut from a piece of Plexiglas (0.5 by 2.25 by 1.5 inch [ea. 12.7 by 57.2 by 38.1 mm]). A 3/16-inch (ca. 4.77-mm) hole was drilled longitudinally in the top of the finger bracket to accommodate a piece of 1/8-inch (ca. 3.2-mm; OD) latex rubber tubing. A vertical hole placed to intercept the longitudinal hole was drilled in the top of the finger bracket. The vertical hole was taped to accommodate an 8-32 brass flow-control screw (Fig. 1, part 2).

The vacuum bulb (Fig. 1, part 8), a Davol (Davol Inc., Providence, R.I.) no. 1765 ounce (ca. 44.4 ml) rubber bulb, was fitted with a Davol 2041 inlet connector and a 2042 tubing connector. These connectors served as one-way valves and are identified as air exhaust and an inlet valve in Fig. 1, parts 7 and 9.

The vacuum bulb was attached to the finger bracket as shown in Fig. 1. A tire patch was bolted to the finger bracket and then cemented (Tech 2-way cement, Technical Rubber Co., Inc., Johnstown, Ohio) to the vacuum bulb.

A piece of glass tubing, 0.4 cm in diameter and 3 cm long, packed with high-efficiency fiber glass-filter media, was inserted in the latex tubing (Fig. 1, part 3) to serve as an in-line filter.

This pipettor functioned quite well with pipettes of any size. To operate, the delivery end of the pipette was placed in the fluid and the opposite end in the pipettor mouth part (Fig. 2). A vacuum was formed by squeezing the vacuum bulb. The air flow was adjusted with the flow-control screw (Fig. 1, part 2). When the desired fluid level was reached, the end of the pipette was transferred from the pipettor to the first joint of the index finger, and fluid flow was controlled as usual.

LITERATURE CITED

1. **Holbert, M. M., D. Vesley, and N. Nick.** 1974. Catalog of pipetting aids and other safety devices for the biomedical laboratory. School of Public Health, University of Minnesota, Minneapolis.
2. **Phillips, G. B., and S. P. Bailey.** 1966. Hazards of mouth pipetting. Am. J. Med. Tech. **32**:127–129.
3. **Songer, J. R., D. T. Braymen, and R. G. Mathis.** 1975. Convenient pipetting station. Appl. Microbiol. **30**:887–888.
4. **Songer, J. R., J. F. Sullivan, and J. W. Monroe.** 1971. Safe, convenient pipetting device. Appl. Micribiol. **21**:1097–1098.
5. **Wedum, A. G.** 1950. Non-automatic pipetting devices for the microbiological laboratory. J. Lab. Clin. Med. **35**:648–651.

Safe, Convenient Pipetting Station

J. R. SONGER,* D. T. BRAYMEN, AND R. G. MATHIS

National Animal Disease Center, Agricultural Research Service, Ames, Iowa 50010

Received for publication 29 July 1975

A simple convenient pipetting station is described that eliminates the need for mouth pipetting. Necessary components, fabrication procedures, and operating techniques are given.

Pipetting by mouth should be discouraged in the laboratory. When working with corrosive or toxic chemicals or infectious disease agents, mouth pipetting is a hazardous procedure. Although mouth pipetting nonhazardous material does not pose a hazard, it becomes so automatic that one often forgets and pipettes hazardous material by mouth. It is a good policy to abstain from putting anything in your mouth while working in a laboratory whether it be cigarettes, food, beverages, or a pipette.

The major hazards of pipetting by mouth are: (i) accidental aspiration of fluid being pipetted, (ii) aspiration of vapors from the fluid being pipetted, (iii) aspiration of aerosols from the fluid when pipetting with unplugged pipette, and (iv) contamination of the proximal end of the pipette by the users contaminated finger, resulting in oral contamination. The hazards are covered in detail by Wedum (3) and Phillips and Bailey (2).

Most commercially available pipetting devices (1) must be attached to the pipette. This makes the pipette more cumbersome and difficult to use. Many pipetting devices will also fit only one size pipette requiring a unit for each size pipette. The unit described in this note (Fig. 1) is a pipetting station that does not attach to the pipette. It will accommodate all sizes of pipettes from the small disposable 1-ml pipette to the large volumetric pipette. The only change necessary when changing pipette sizes is the level of vacuum.

Fabrication procedure. The pipetting station base was cut from a piece of brass plate 1 inch (2.54 cm) thick (Fig. 2, no. 4). A hole (0.837 inch in diameter [ca. 2.13 cm] by 0.75 inch deep [ca. 1.9 cm]) was machined in the center of the base. The stand pipe (Fig. 2, no. 3) was threaded on one end, 14 threads per inch. The pipe die was adjusted to cut the threads deeper than normal to accommodate the coarser threads of the coupling. The other end of the stand pipe was press fitted into the base. A 11/32 inch in diameter (ca. 0.88 cm) hole was drilled from the side of the base to converge with the standpipe (Fig. 2). One-eighth inch (ca. 0.311 cm) NPT threads were tapped in this hole to accommodate the needle valve.

The pipetter head (Fig. 2, no. 8), 7/8 inch (ca. 2.219 cm) in diameter and 5/16 inch (ca. 0.8 cm) thick, was machined from brass plate. The head was machined to securely hold a size B vaccine bottle stopper (Fig. 2, no. 10). A hole was drilled in the side of the pipetter head to accommodate the tip of the flexible spout. The pipetter head was then silver soldered to the flexible spout tip. A hole (1/32 inch [ca. 0.08 cm] in diameter) was drilled through the center of a size B vaccine bottle stopper (Fig. 2, no. 9) with a small hand grinder that had a carborundum bit.

The stopper was inserted into the head, the spout was connected to the standpipe, and the

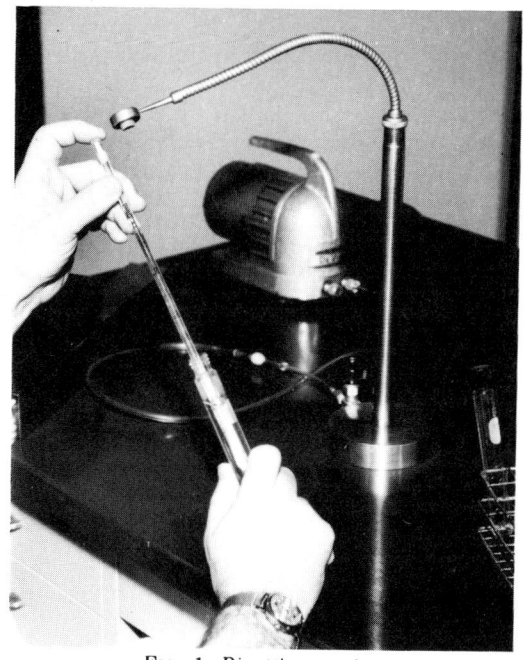

FIG. 1. *Pipetting station.*

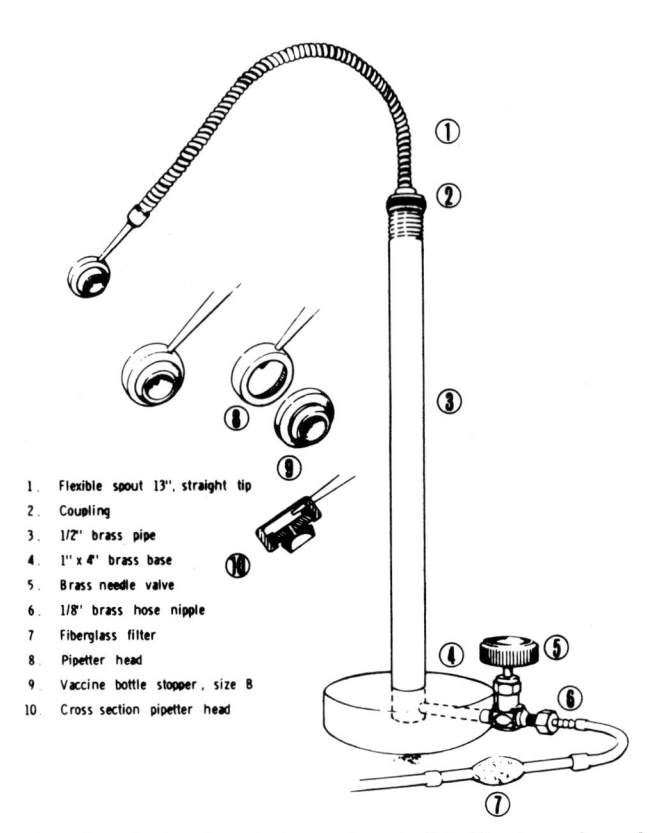

1. Flexible spout 13", straight tip
2. Coupling
3. 1/2" brass pipe
4. 1" x 4" brass base
5. Brass needle valve
6. 1/8" brass hose nipple
7. Fiberglass filter
8. Pipetter head
9. Vaccine bottle stopper, size B
10. Cross section pipetter head

FIG. 2. *Construction drawing of pipetting station and parts list. The brass base, brass standpipe, and pipetter head were made in a machine shop. The filter was made from the bulb section of a 5-ml volumetric pipette filled with FM004 filter media (Owens-Corning Fiberglas Corp.). The other parts were obtained as follows: (i) straight tip flexible spout (No. 328) (13 inches [ca. 33.02 cm] long), Eagle Mfg. Co.; (ii) coupling (HA-24), Eagle Mfg. Co.; (iii) gasket (HA-10), Eagle Mfg. Co.; (iv) brass needle value (M-312M2B) ($^{1}/_{8}$-inch NPT-M inlet and outlet), Matheson Gas Products; (v) brass hose nipple ($^{1}/_{8}$-inch NPT-F), McMaster-Carr Supply Co.*

valve, hose nipple, and filter were attached to the base. A small diameter hose connected the filter to a low vacuum source (Fig. 1).

Use of pipetting station. The pipetting station is very easy to use (Fig. 1). One simply holds the pipette as usual, between the thumb and second finger. The proximal end of the pipette is placed into the concave opening of the vaccine bottle stopper and, at the same time, the index finger is pressed down upon the top of the pipetter head. When the fluid reaches the desired level in the pipette, pressure on the pipetter head is released. At the same time the pipette is withdrawn and the index finger is placed over the end as is done when pipetting by mouth. From this point on pipetting is the same as when pipetting by mouth.

With a little practice pipetting can be done as fast with this pipetting station as by mouth, or even faster. If the pipetting station becomes contaminated with pipetting liquids, it can be disconnected from the vacuum source and sterilized in an autoclave.

LITERATURE CITED

1. Holbert, M. M., D. Vesley, and N. Vick. 1974. Catalog of pipetting aids and other safety devices for the biomedical laboratory. School of Public Health, University of Minnesota, Minneapolis.
2. Phillips, G. B., and S. P. Bailey. 1966. Hazards of mouth pipetting. Am. J. Med. Tech. 32:127–129.
3. Wedum, A. G. 1950. Non-automatic pipetting devices for the microbiological laboratory. J. Lab. Clin. Med. 35:648–651.

Safety Equipment

Self-Sterilizing Inoculating Loop

J. R. SONGER, J. L. RILEY, AND D. T. BRAYMEN

National Animal Disease Laboratory, Veterinary Sciences Research Division, U.S. Department of Agriculture, Ames, Iowa 50010

Received for publication 8 March 1971

A self-sterilizing inoculating loop consisting of a step-down transformer and an adjustable timing circuit are described.

Use of an open flame for sterilizing inoculating loops is sometimes unsafe, inconvenient, or impossible. Some anaerobic studies require an atmosphere that will not support a flame (1). Heat buildup is often encountered when open flames are used in recirculating biohazards hoods. The self-sterilizing inoculating loop described in this note was developed to help eliminate these problems (Fig. 1).

Trotman and Drasar (2) modified a cautery burner for use as a self-sterilizing inoculating loop. This loop was used to transfer anaerobic bacteria in a nitrogen-filled anaerobic cabinet. Trotman (1) also described a similar self-sterilizing loop used in an automatic plate streaker.

We found that a current of 6 v a-c flowing through an 8-inch (20.3 cm) piece of size 24 chromel "A" wire with a resistance of 1.062 ohms per ft was adequate to heat the wire to a glowing red in 10 sec.

The control box (Fig. 1) consists of a step-down transformer for supplying power and a timing circuit for controlling the time power is applied to the inoculating loop. Construction details are shown in Fig. 2 and 3. By pressing the switch (S_2), the relay (K_2) is locked closed through the time delay relay (K_1). After 10 sec, the time-delay relay contacts open, releasing the relay (K_2), which turns off power to the inoculating loop.

The parts and their sources are as follows: S_1, 120-v on-off push canopy switch with leads, Leviton no. 579; S_2, pushbutton momentary contact switch, Switchcraft type 913 56A 5418; I_1,

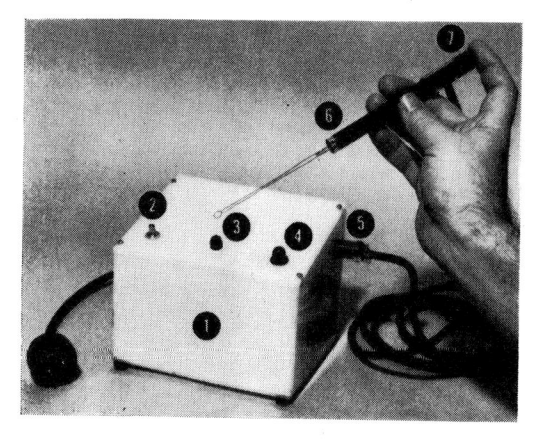

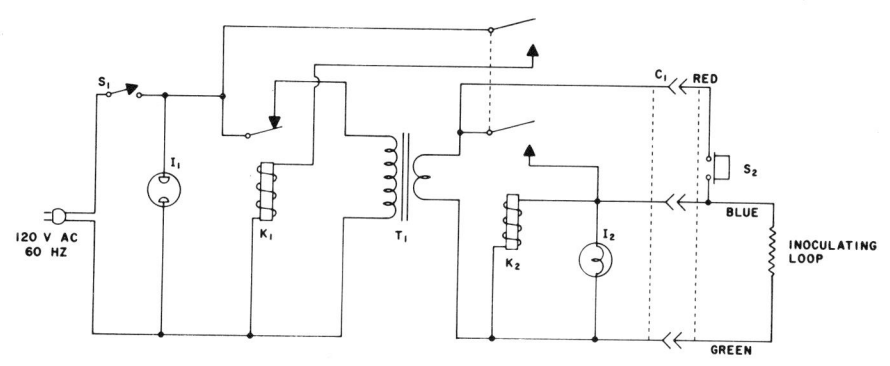

FIG. 1. *Self-sterilizing inoculating loop. (1) Control box, (2) on-off switch (S1), (3) on-off switch indicator light (I1), (4) inoculating loop switch indicator light (I2), (5) connector (C1), (6) Inoculating loop, and (7) inoculating loop pushbutton switch.*

FIG. 2. *Schematic diagram of the self-sterilizing inoculating loop.*

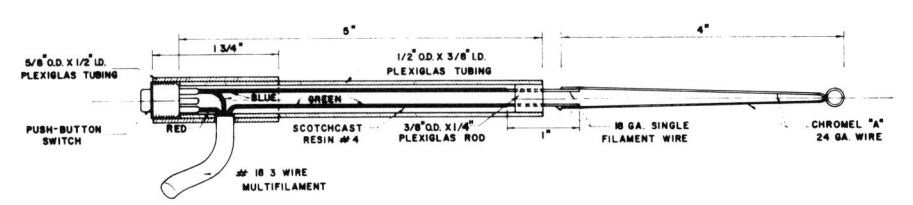

FIG. 3. *Cross section view of the self-sterilizing inoculating loop.*

pilot lamp 120-v neon, Leecraft Snaplite 32-2111; I$_2$, pilot lamp 6.3-v a-c, Drake LH 31; K$_1$, time delay relay 120-v a-c, solid state timer, model SS 06262 JJ, Intermatic Time Control, International Register Co.; T$_1$, filament transformer 6.3-v a-c 10 amps, Stancor P6464; K$_2$, relay 6.3-v a-c 10 amp, Potter and Brumfield CHB 38-70003; C$_1$, connector, amphenol connector female 126011, male 126217; inoculating loop, Chomel "A" size 24 wire, 8 inches in length with a resistance of 1.062 ohms per ft, Ogden Mfg. Co.; Scotchcast resin No. 4, Minnesota Mining and Mfg. Co.; Plexiglas 1.5 ft^2 of 0.25 inch white, Rhome and Haas Co.

The inoculating loop is readily adaptable for use in a confined space or controlled atmosphere cabinet. With the amphenol connector located in the cabinet wall, the control box can be located outside the cabinet.

A solution containing 3.4×10^8 *Bacillus subtilis* spores per ml was used in evaluating this loop. The loop was dipped into the spore solution. It was removed from the solution, heated, and used to streak a nutrient agar plate. This procedure was repeated 10 times. No colonies appeared in any of the 10 plates.

The total cost of material for construction of the inoculating loop was approximately $35.

LITERATURE CITED

1. Trotman, R. E. 1970. *In* Ann Braillie and R. J. Gilbert (ed.) Automatic methods in diagnostic bacteriology in automation, mechanization and data handling in microbiology, p. 211–221. Academic Press Inc., New York.
2. Trotman, R. E., and B. S. Drasar. 1968. Electrically heated inoculating loop. J. Clin. Pathol. 21:224-225.

First Aid

WHAT IS FIRST AID?

First aid is the immediate care given to a person who has been injured or has been suddenly taken ill. It includes well-selected words of encouragement, evidence of willingness to help, and promotion of confidence by demonstration of competence.

In case of serious injury or sudden illness, while help is being summoned, immediate attention should be given to these first aid priorities:

1. Effect a prompt rescue (for example, remove an accident victim from a room containing carbon monoxide or noxious fumes).
2. Ensure that the victim has an open airway and give mouth-to-mouth artificial respiration, if necessary.
3. Control severe bleeding.
4. Give first aid for poisoning, or ingestion of harmful chemicals.

Above all, know the limits of your capabilities and make every effort to avoid further injury to the victim in your attempts to provide the best possible emergency first aid care.

It is recommended that teachers know about potentially dangerous medical health problems students may have.

Courses Available
American Red Cross:
    Standard First Aid and Personal Safety   14 hours
    Standard First Aid--Multimedia System   7½ hours
    CPR--CardioPulmonary Resuscitation   6 hours
Instructor Courses:
    Standard First Aid Multimedia System   1 hour
    Orientation on CPR   2½ hours
Contact:
    American Red Cross, Iowa Division
    2116 Grand Avenue, Des Moines, Iowa 50312
    Phone: (515) 243-7681

Much of the material for this section was derived, with permission, from: *Standard First Aid and Personal Safety*, American National Red Cross. 1973, Doubleday and Co., Garden City, New York.

Iowa Heart Association
    CPR--CardioPulmonary Resuscitation
        Heart Saver Course   4 hours
        Basic Rescuer Course   6-8 hours
Instructor Courses:
    CPR - Contact Association for information
Contact:
    Iowa Heart Association
    3810 Ingersoll, Des Moines, Iowa 50312
    Phone: (800) 362-2440

<u>Traumatic Shock</u>

    Traumatic shock may result from serious injuries of all kinds, hemorrhage or loss of body fluids, heart attack, burns, poisoning by chemicals, gases, alcohol, or drugs. Shock also results from lack of oxygen caused by obstruction of air passages or injury to the respiratory system.

    Shock is aggravated by pain, by rough handling, and by delay in treatment and, if untreated, can result in death to the victim.

    Symptoms include: pale, cold clammy skin; weakness; rapid pulse, shallow and rapid breathing; and nausea.

Procedures:
1. Keep victims lying down.
2. Cover victims only enough to prevent loss of body heat.
3. Get medical help as soon as possible.
4. Victims who are having difficulty breathing may be placed on their backs with head and shoulders raised.
5. Victims may improve if the feet are raised from 8 to 12 inches. Lower the feet back to a level position if any difficulty in breathing or additional pain results.
6. If medical attention will be delayed for over one-half hour, small amounts of fluids may be given a victim who is conscious. Lukewarm water is usually the most available.

Further Reference: (See Transportation.)

<u>Artificial Respiration</u>

    The objectives of artificial respiration are to maintain an open airway through the mouth, nose, or stoma and to restore breathing by maintaining an alternating increase and decrease in the expansion of the chest.

    Recovery is usually rapid except in cases involving carbon monoxide poisoning, drug overdose, or electrical shock.

    Artificial respiration should be continued until victims start to breathe for themselves; medical authority pronounces them dead; they are dead beyond any doubt.

    The mouth-to-mouth method:

1. Clear any foreign matter from the mouth.
2. Tilt the victim's head back to open the air passage. Make sure the chin is pointing upward.
3. Pinch the victim's nostrils shut with thumb and index finger.
4. Blow air into the victim's mouth. Your mouth should be open wide and sealed tightly around the mouth of the victim.
5. Stop blowing when the victim's chest is expanded. Remove your mouth. Turn your head and listen for the victim's exhalation. Watch the chest to see that it falls.
6. Repeat the cycle 12 times per minute (every 5 seconds).
7. Get medical assistance.

First Aid 133

Further Reference: (See Shock, Transportation.)

Heimlich Maneuver

An effective way to remove an object blocking the windpipe is the Heimlich maneuver. To perform this maneuver:

1. Stand behind the victim and place your arms around the victim's waist.
2. Make a fist and place it so that the thumb is against the victim's abdomen slightly above the navel and below the ribcage.
3. Grasp your fist with the other hand and then press your fist into the victim's abdomen with a quick upward thrust.
4. This action forces air out of the victim's lungs and blows the object from the windpipe.

If the victim has collapsed or is too large for you to support or place your arms around:

1. Lay the victim down, face up.
2. Face the victim; kneel, straddling the victim's hips.
3. Place one of your hands over the other, with the heel of the bottom hand on the victim's abdomen, slightly above the navel, and below the ribcage.
4. Press your hands into the victim's abdomen with a quick upward thrust.

When applying the Heimlich maneuver, be careful not to apply pressure on the victim's ribs. Such pressure may break the ribs of a child or an adult.

Red Cross First Aid Procedures for Choking
CONSCIOUS VICTIM:
   The following sequence of maneuvers should be performed immediately and in rapid succession on the conscious victim in the sitting, standing, or lying position:

1. *Back Blows*--four in rapid succession. If ineffective proceed to 2.
2. *Manual Thrusts*--eight in rapid succession. Proceed to 3.
3. Repeat back blows and manual thrusts until they are effective or until the victim becomes unconscious.

UNCONSCIOUS VICTIM:
   If the victim is not breathing but *can* be ventilated, proceed with mouth-to-mouth resuscitation.
   If the victim is not breathing and *cannot* be ventilated quickly perform the following sequence:

1. *Back Blows*--four in rapid succession. If ineffective proceed to 2.
2. *Manual Thrusts*--eight in rapid succession with victim lying on back. If ineffective proceed to 3.
3. *Finger Probe*--Use your index finger and probe deep into the throat using your other hand to hold the mouth open wide. If unsuccessful proceed to 4.
4. Repeat sequence--persist.

Transportation of a Victim

It should be recognized that more harm can be done through improper rescue and transportation than through any other measures associated with emergency assistance.

Unless there is immediate danger to life, victims should not be transported until such life-threatening problems as airway obstruction and hemorrhage are cared for and wounds are dressed.

Most schools are located where rescue services, ambulances, and other medical support services are readily available. The wise first aider will use those services in preference to using makeshift transportation procedures.

Animal Bites

Animal bites may cause punctures, lacerations, or avulsions. Not only do the wounds need care but consideration must be given to the possibilities of infection, including rabies and tetanus.

After the bite:

1. Keep the animal alive and make every effort to restrain the animal for observation.
2. If it is necessary to kill the animal, take precautions to keep the head free from damage.
3. Arrange for medical attention to the victim.

Procedures:
1. Immobilize the affected body part.
2. Wash the wound and the area around it with soap and water, flush the bitten area and apply a dressing.
3. Treat for traumatic shock.

Further Reference: (See Wounds, Traumatic Shock.)

Blisters

Blisters can result from friction or from burns. A broken blister must always be treated as an open wound.

Procedures:
1. If all pressure can be relieved until the fluid is absorbed, blisters are best left unbroken.
2. Self-care for blisters should not be attempted when the blister fluid lies deep in the palm of the hand or sole of the foot.
3. Blisters resulting from burns should never be broken.
4. If friction blisters must be broken:
   a. Wash the entire area with soap and water.
   b. Make a small puncture hole at the base of the blister with a sterile needle.
   c. Apply a sterile dressing and protect the area from further irritation.

Further Reference: (See Wounds.)

Eye Injuries

Injury to the eyelid is much like other soft tissue injuries, and the first aid treatment is similar.

A blunt injury, or contusion, often occurs from a severe direct blow as from a fist or in an explosion.

Penetrating injuries of the eye are extremely serious. If an object lacerates or penetrates the eyeball, a loss of vision or even blindness can result.

Procedures:
1. Injury to the eyelid

First Aid

   a. Stop hemorrhage by gently applying direct pressure.
   b. Apply a sterile or clean dressing.
   c. Seek medical assistance without delay.
2. Blunt injury or contusion
   a. In serious cases, the structure of the eye may be torn or ruptured.
   b. Secondary damage may be produced by the effects of hemorrhage, and later by infection.
   c. Apply a dry sterile or clean dressing.
   d. Transport to the hospital with the victim lying flat.
3. Penetrating injuries of the eye
   a. Make no attempt to remove the object or to wash the eye.
   b. Cover both eyes to lessen eye movement. Avoid making the covering so tight as to put pressure on the affected eye.
   c. Keep the victim quiet.
   d. Transport to the hospital immediately with victim lying down, face up.

Further Reference: (See Transportation, Traumatic Shock, Wounds.)

Respiratory Emergencies

A respiratory emergency is one in which normal breathing stops or in which breathing is so reduced that oxygen intake is insufficient to support life. Symptoms include:

1. Breathing movements have stopped.
2. The victim's tongue, lips, and fingernail beds become blue.
3. There is a lack of consciousness.
4. The pupils of the eye become dilated.

Procedures:
Carbon monoxide poisoning:
   a. Ventilate area before attempting rescue.
   b. Remove victim to fresh air.
   c. The victim's lips and skin may be cherry red.
   d. Give all necessary first aid, including artificial respiration.
   e. Recovery may be slow. It is often necessary to continue artificial respiration for a long time.
   f. Treat for traumatic shock.
   g. Get medical assistance as soon as possible.

Choking--foreign body preventing breathing:
   a. Try to manually clear the object from the victim's mouth using a sweeping motion of the index and middle finger. Insert them inside the victim's cheek. Slide them deeply into the throat to the base of the tongue, and out along the inside of the other cheek.
   b. If breathing does not start immediately, carry out the Heimlich maneuver or Red Cross First Aid procedures.
   c. Treat for traumatic shock.
   d. Get medical assistance if needed.

Electrocution:
   a. Cut off the power at the main switch before attempting a rescue.
   b. Because electricity tends to paralyze muscles used in breathing, recovery time may be prolonged. Continue artificial respiration for a long time.
   c. After breathing starts, treat for traumatic shock.
   d. Care for any burns that may have resulted from the electrical shock.

Further References: (See Artificial Respiration, Traumatic Shock, Heimlich Maneuver, Red Cross Choking Procedures, Burns.)

Wounds

A wound is a break in the continuity of the tissues of the body, either internal or external.

First aid for wounds includes stopping bleeding immediately, protecting against infection, providing care to prevent traumatic shock, and obtaining medical assistance if needed.

The loss of blood can cause traumatic shock. Loss of as little as one quart of blood can result in loss of consciousness of the victim. Because it is possible for a victim to bleed to death in a very short period of time, any large, rapid loss of blood should be stopped immediately.

Procedures:
Techniques to stop severe bleeding:
1. Direct pressure
   Place a thick pad of cloth over the wound. Apply direct pressure by placing the palm of the hand over the dressing. The pad will absorb the blood and allow it to clot while the pressure slows the flow. (Do not disturb blood clots. If blood soaks entire pad, place another pad on top of the first.)
2. Elevation
   Unless there is evidence of a fracture, a severely bleeding open wound of hand, neck, arm, or leg should be elevated. Elevation uses gravity to help reduce blood pressure in the injured area and thus aids in slowing down the loss of blood. Direct pressure must be continued.
3. Pressure on the supplying artery.
   a. The use of the pressure point technique temporarily compresses the main artery supplying blood to the affected part against the bone.
   b. If the use of a pressure point should be necessary, do not substitute its use for direct pressure and elevation, but use the pressure point *in addition to* those techniques.
   c. Use the brachial artery for the control of severe bleeding from an arm wound.
   d. Use the femoral artery for the control of severe bleeding from a leg wound.
4. Tourniquet
   The use of a tourniquet is dangerous and should be a last resort. The decision to apply a tourniquet is in reality a decision to risk the sacrifice of a limb in order to save a life.
   a. The tourniquet should be at least two inches wide. Place the tourniquet just above the wound.
   b. Wrap the tourniquet band tightly around the limb twice and tie a half knot.
   c. Place a short, strong stick or similar object, on the half knot and tie two overhand knots on top of the stick.
   d. Twist the stick to tighten the tourniquet until bleeding stops.
   e. Secure the stick in place.
   f. Make a written note of the location of the tourniquet and the time it was applied and attach the note to the victim's clothing.
   g. The tourniquet should not be loosened except on the advice of a physician.
   h. The tourniquet should never be covered.

First Aid

Prevention of contamination and infection:
1. Wounds with serious bleeding
   a. The pad initially placed on the wound should not be removed or disturbed.
   b. Cover the wound and pad with a clean dressing.
   c. Transport the victim to medical assistance.
2. Wounds without severe bleeding
   a. Apply a dry sterile bandage or clean dressing and secure it firmly in place.
   b. Caution the victim to see a physician promptly if evidence of infection appears.

Further Reference: (See Traumatic Shock, Transportation.)

Sudden Illness

First aiders often encounter emergencies that are not related to injury but arise from either sudden illness or a crisis in a chronic illness. Unless the illness is minor and brief, such as a fainting attack, nosebleed, or a headache, medical assistance should be sought.

Many persons suffering from heart disease, apoplexy, epilepsy, or diabetes carry an identification card or bracelet that contains information about the type of illness and the steps to be followed if the persons are found unconscious. Search the victim (in the presence of witnesses) for such identification.

Heart attack
A. Signs and symptoms
   1. Persistent chest pain, usually under the sternum (breastbone).
   2. Gasping and shortness of breath.
   3. Extreme pallor or bluish discoloration of the lips, skin, and fingernail beds.
   4. Extreme prostration.
   5. Shock.
   6. Swelling of the ankles.
B. First aid
   1. Place the victim in a comfortable position, usually sitting up.
   2. If the victim is not breathing, begin artificial respiration.
   3. Treat for shock.
   4. Get medical assistance.
   5. Seek medical advice before transporting the victim.

Fainting
A. Signs and symptoms
   1. Extreme paleness.
   2. Sweating.
   3. Coldness of the skin.
   4. Dizziness.
   5. Nausea.
B. First aid
   1. Leave the victim lying down.
   2. Loosen tight clothing.
   3. Maintain an open airway.
   4. Give no liquids.
   5. Unless recovery is prompt, seek medical assistance.

Epilepsy
- A. Signs and symptoms
    1. Repeated convulsions -- "grand mal" seizures.
    2. A milder form of epilepsy occurs without convulsions. There may be only brief twitching of muscles, "petit mal" seizures, and momentary loss of contact with the surroundings ("grand mal" seizures are more common).
- B. First aid
    1. Push away objects--do not restrain the victim.
    2. Do not force a blunt object between the victim's teeth.
    3. When jerking is over, loosen clothing around the victim's neck.
    4. Keep the victim lying down.
    5. Keep an open airway.
    6. If breathing stops, give artificial respiration.
    7. After the seizure, allow the victim to sleep or rest.
    8. If convulsions occur again, get medical assistance.

Burns

A burn is an injury that results from heat, chemical agents, or radiation. It may vary in depth, size, and severity.

The object of first aid for burns is to relieve pain, prevent contamination, and treat for traumatic shock.

Procedures:
1. *Thermal burns--first and second degree*
    a. Are identified by one or more of the following symptoms: redness or discoloration, mild swelling or pain, development of blisters, or wet appearance of the surface of the skin.
    b. Apply cold water applications, or submerge the burned area in cold water.
    c. Apply a dry dressing if necessary. It will almost always be necessary in cases involving second degree burns.
    d. Blisters should not be broken or tissue removed.
    e. Preparations, sprays, or ointments should not be used on burns.
    f. Treat for traumatic shock if necessary.
2. *Thermal burns--third degree*
    a. The usual signs are deep tissue destruction, white or charred appearance, and complete loss of all layers of the skin.
    b. Cover burns with thick, sterile dressings.
    c. If hands are involved, keep them above level of the victim's heart.
    d. Keep burned feet or legs elevated.
    e. A cold pack may be applied to the face, hands, or feet but extensive burn areas should not be immersed.
    f. Adhered particles of charred clothing should be left in place.
    g. Ointments, sprays, or other preparations should not be applied to burned areas.
    h. Arrange for medical assistance and transportation to the hospital.
3. *Chemical burns of the skin*
    a. Wash away the chemical with large amounts of water, using a shower or hose, as quickly as possible and for at least 5 minutes. Remove clothing from affected areas.
    b. If first aid directions for burns caused by specific chemicals are available, follow them after the flushing with water.
    c. Apply necessary dressings and get medical assistance.

First Aid

4. *Burns of the eye*
    a. Acid burns
        1) Thoroughly wash the face, eyelids and eye for at least 5 minutes. If the victim is lying down, turn the head to the side, hold the eyelid open and pour water from the inner corner of the eye outward.
        2) If a weak soda solution (1 teaspoon of baking soda added to 1 quart of water) can be made quickly, use the solution after the first washing of the eye with tap water.
        3) Cover the eye with a dry, clean protective dressing. Cotton should not be used.
        4) Get medical assistance.
        5) Caution the victim against rubbing the eye.
    b. Alkali burns
        1) Flood the eye thoroughly with water for 15 minutes.
        2) If the victim is lying down, turn the head to the side. Hold the eyelid open and pour the water from the inner corner outward.
        3) Remove any loose particles of dry chemical floating on the surface of the eye by lifting them off gently with sterile gauze.
        4) Irrigation of the eye is not appropriate.
        5) Immobilize the eye by covering it with a dry pad or dressing.
        6) Get medical assistance.

Further Reference: (See Traumatic Shock, Wounds, Blisters, Transportation.)

Poisoning

A poison is any substance, solid, liquid, or gas, that tends to impair health, or cause death, when introduced into the body or on to the skin surface.

The objectives in treatment of poisoning by mouth are to dilute or neutralize the poison as quickly as possible, to induce vomiting (except when corrosive poisons are swallowed or if victim is unconscious or having convulsions), to maintain respiration, to preserve vital functions, and to seek medical assistance without delay.

Procedures:
Poisoning by mouth
1. General:
    a. If victim is unconscious, keep the airway open, administer artificial respiration if necessary, and transport as soon as possible to obtain medical assistance.
    b. While you are giving first aid to the victim, have someone else get advice by telephone from a doctor, hospital, or poison control center.
    c. If vomiting is to be induced, do so by tickling the back of the victim's throat or by giving a nauseating fluid such as syrup of ipecac or mustard and water.
2. If you *do not* know what poison the victim swallowed:
    a. Dilute the poison with water or milk.
    b. Try to find out what poison has been swallowed.
    d. Get medical assistance immediately.
3. If you *do* know that the victim has *not* swallowed a strong acid, strong alkali, or petroleum product, but do not have the original container:
    a. Dilute the poison with water or milk.
    b. Induce vomiting.
    c. Get medical assistance immediately.

4. You have the original container from which the known poison came:
   a. If a specific antidote is described on the label of a commercial product, administer it according to directions if the victim is conscious.
   b. Save the label or container.
   c. Get medical assistance immediately.
5. If a strong *acid* is involved:
   a. Dilute with one glass of water or milk.
   b. Neutralize with milk of magnesia or other weak alkali, mixed with water--3 or 4 glasses for adults or 1 to 2 glasses for children.
   c. You may find it necessary to administer milk, olive oil, or egg white as a demulcent to coat and soothe the stomach and intestines.
   d. Get medical assistance immediately.
6. If a strong *alkali* is involved:
   a. Dilute with one glass of water or milk.
   b. Neutralize with vinegar or lemon juice, mixed with water--3 or 4 glasses for adults or 1 to 2 glasses for children.
   c. You may find it necessary to administer milk, olive oil, or egg white as a demulcent to coat and soothe the stomach and intestines.
   d. Get medical assistance immediately.
7. If a petroleum product is involved:
   a. Dilute with one glass of water or milk.
   b. Get medical assistance immediately.

Contact with poisonous plants
1. The majority of skin reactions following contact with offending plants are allergic in nature and are characterized by the general symptoms of headache and fever, itching, redness, and a rash. Ordinarily, the rash begins within a few hours after exposure, but it may be delayed for 24 to 48 hours.
2. Remove contaminated clothing; wash all exposed areas with soap and water, followed by rubbing alcohol.
3. Apply calamine or other soothing skin lotion if the rash is mild.
4. Seek medical assistance if a severe reaction occurs, or if there is a known history of sensitivity.

Poisoning by marine life
1. A variety of species of fish are equipped with venom apparatus attached to dorsal or other spines. Examples include catfish, weever fish, scorpion fish (including zebrafish), toadfish and surgeonfish. First aid treatment relates to the symptoms, since little is known regarding antidotes.
2. Stings from jellyfish and the Portuguese man-of-war produces a venom which in turn produces burning pain, a rash with minute hemorrhages in the skin, shock, muscular cramping, nausea, and respiratory difficulty.
   a. Wipe off affected area with a towel, and wash the area thoroughly with diluted ammonia or rubbing alcohol.
   b. Give aspirin for pain.
   c. Seek medical assistance if symptoms are severe.

Poisoning by insects
1. Stings from ants, bees, wasps, hornets, and yellow jackets can cause death due to allergic reaction.
2. Spiders in the U.S. are generally harmless. However, black widow spiders and the brown recluse or violin spider are poisonous and bites can be harmful. Medical assistance should be sought.
3. Scorpions inject venom through a stinger in the tail. Fatalities have been recorded. Medical assistance is needed.

When severe reactions result from any insect bite:
1. Give artificial respiration if indicated.

2. Apply a constricting band above the injection site on an arm or leg. You should be able to slip your index finger under the band when it is in place. If not, loosen the band. (The band should be removed after 30 minutes.)
3. Keep the affected part down, below the level of the victim's heart.
4. Apply ice contained in a towel or plastic bag, or cold cloths, to the site of the sting or bite.
5. Give aspirin for pain.

Poisoning by venomous snakes
1. Keep the victim calm and quiet. Transport victim to a source of medical assistance as quickly as possible.
2. Immobilize the arm or leg in a lowered position, keeping the involved area below the level of the victim's heart.
3. Treat the victim for shock.
4. Give artificial respiration if needed.
5. Get immediate medical assistance.

Further Reference: (See Artificial Respiration, Traumatic Shock, Transportation, Wounds.)

INDEX

Accidents
  all activities, 6
  in classroom, 6
  data, general, 5-7
  in science, nature of, 7
  severity of, 6
Accident/incident reporting, 101-3
  checklist, 113
American National Standards Institute (ANSI), 19, 21, 115
Animal bites, 7, 34, 134, 140-41

Behavior. *See* Laboratory behavior
Biology, 25
  animals, 34, 89, 134
  aquaria, 28
  bites, animal, 134, 140-41
  blood sampling, 38
  cages, animal, 27, 89
  chemicals, dangerous, 39, 49, 50, 63
  disposal, 27
  dissections, 37
  extractions, 38
  facilities, 26, 88
  field activities, 29, 81-84
  killing jars, 35
  labeling, 27, 68
  microorganisms, 28
  pipetting, 28, 125, 127
  plant room, 89
  plants, poisonous, 29, 30-31, 34, 33, 140
  radioactive materials, 40, 47, 71, 78
  refrigerators, 26, 51, 116-21
  specimens, preserved, 26, 36, 83
  sterilization, 28, 129
  storage, 26, 88
  terraria, 28
  checklist, 108
Blind (visually impaired). *See* Handicapped students
Burns
  acid/alkali, 138-39

Checklists. *See* Appendix A, 105-13
Chemicals
  corrosive, hazardous, reactive, 45, 47, 56
  dispensing, 59
  disposal, 65, 67
  handling, 54
  incompatible, 56, 57, 59
  labeling of, 68
  research in, 58
  spills, 66
  storage, 41, 51-52, 124
Chemistry, 41
  carcinogens, 49, 50, 63
  compressed gases, 48, 53
  experimental procedures, 55
  explosives, 48
  facilities, 89
  flammable liquids and solvents, 42, 44-46, 51, 124
  inventory control, 52
  lasers, 75, 78, 115
  purchasing, suggestions, 52
  radioactive materials, 51, 71-75, 78
  refrigerators, 51, 116
  safety inspections, 53
  X rays, 75, 78
  checklist, 110
CHEMTREC, 67
Class size, 25, 87-88
Communications, 4
Contact lenses, 19, 21, 24

Deaf (hearing impaired). *See* Handicapped students
Disposal. *See* Biology, Chemicals, and Physics

Emotionally disturbed/disruptive students, 97
Employer-employee responsibilities, Code of Iowa, 12-13
Explosion-safe refrigerators, 51, 116
  modification of home refrigerator, 116-21
Eyes, 21
  burns of, 139
  care, 21
  emergency equipment, 23
  injuries, 23, 134, 139
  protective devices for, 12, 19, 114
    approval standards (Z87.1-1979), 19
    cleaning and care, 20
    face shields, 19, 115
    goggles, 19, 115
    laser spectacles, 19, 115
    spectacles, 19, 115
    visitor glasses, 20
  wash for, 123
  checklist, 107, 108

Field activities, 29, 81
  collections, 26, 36, 83-84
  consent forms, 81-82
  pre-trip site visits, 83
  supervision, 82
  transportation, 83
  checklist, 112
Fire fighting equipment, 92-93
First aid, 131
  artificial respiration, 132
  bites, animal, 7, 34, 134, 140-41
  burns, blisters, 77, 134, 138
  courses, 131
  choking: Heimlich and Red Cross maneuvers, 133
  heart attack, 137
  epilepsy, 137
  eye burns, 138-39
  eye injuries, 23, 134, 139
  fainting, 137
  poisoning, 139
  respiratory emergencies, 135
  shock (traumatic), 132
  transporting victim, 133
  wounds, 5, 7, 77, 136
Flammable liquid cabinet (construction), 124
Freezing (eryogenic), 78

Good Samaritan Law, 11

Handicapped students, 95
  emotionally disturbed/disruptive, 97
  hearing impaired, 96
  guidelines for, 96
  learning disabled, 97
  mentally retarded, 97
  physically impaired, 97
  PL 94-142, 95
  speech impaired, 97
  visually impaired, 96
  Vocational Rehabilitation Act, 95
  checklist, 113

Health hazards, 60, 62

Iowa Department of Environmental Quality, 67

Laboratory behavior, 25, 54
Lasers. *See* Chemistry, Physics
Learning disabled students. *See* Handicapped students
Legal liability, 11
  case studies, 14
    decisions against instructor, 14
    decisions favorable to instructor, 16
  duties of teacher in, 10
  insurance, 12
  checklist, 107

Mentally retarded students. *See* Handicapped students

Negligence, 9
  avoidance of, 13
  in tort law, 9
NFPA hazard diagram, 69

Pathogens, 28
Physically impaired students. *See* Handicapped students
Physical plant and facilities, 87
  animal room, 89
  calculating teaching stations, 88
  classroom size, 25, 88
  electrical equipment, 92
  enrollment, 87-88
  fire fighting equipment, 92

Index 145

furniture and equipment, 91
gas service lines, 92
laboratory facilities, 88-90
multipurpose science room, 90
plant room, 89
plumbing, 91
ventilation, 91
checklist, 112
Physics, 47
electrical, 78, 92
facilities, 90
lasers, 79
radiation, 78
vacuum tubes, 77
X-ray sources, 79
checklist, 111
Poisoning, 139
by insects, 140
by marine life, 140
by plants, 29-34, 140
by snakes, 141

Safety components, 3

School Laws of Iowa, 9, 11
due care, 9
Schools
plant and personnel services, 3
safety programs, 3, 7
Sovereign Immunity Doctrine, 12
Speech impaired students. *See* Handicapped students
Student research, 85
home study, 86
project design, 85
supervision of, 85
checklist, 112

Tort law, 9
Toxicity, 61-62

Ventilation, 91

X rays. *See* Chemistry, Physics

Z87 logo, 19